Along the Huron

A GUIDE TO

THE NATURAL COMMUNITIES OF

THE HURON RIVER CORRIDOR

IN ANN ARBOR, MICHIGAN

Parks&Recreation
CITY OF ANN ARBOR

NATURAL AREA PRESERVATION DIVISION

Along the Huron: The Natural Communities of the Huron River Corridor in Ann Arbor, Michigan

Parks&Recreation
CITY OF ANN ARBOR

Natural Area
Preservation
Division

Department of Parks and Recreation
Natural Area Preservation Division
1831 Traver Road
Ann Arbor, MI 48105

Phone: 734-996-3266 or 734-994-4834
Fax: 734-997-1072

We welcome your thoughts and comments on this publication. Please contact the
Natural Area Preservation office at the above number with any questions or ideas.

> *The mission of the Natural Area Preservation Division is to protect, restore, and*
> *champion Ann Arbor's natural areas, especially those in city parks.*

ACKNOWLEDGMENTS

This project was made possible by citizens who contributed to the Nongame Wildlife Fund on the Michigan income tax form or who gave a direct donation to the Fund. These sources provided funding for the production of this guide.

This guide would not have been possible without the cooperation of staff from the University of Michigan and the Washtenaw County Department of Parks and Recreation, who own and/or manage two of the natural areas included in this guide. Specifically, we would like to thank Guy Smith and Liz Elling, Nichols Arboretum staff, and Professors Bob Grese and Harry Morton of the University of Michigan, School of Natural Resources and the Environment (SNRE) for their assistance with the Nichols Arboretum section. Special thanks also to Dr. Andrew Brenner and the SNRE Geographical Information Systems (GIS) lab for help with GIS mapping. Also thanks to Matt Heumann for his help with the section on Parker Mill-Forest Park which is run by the Washtenaw County Parks and Recreation Commission.

Special thanks goes to the volunteers who assisted with inventories of plants, amphibians, butterflies, and breeding birds along the Huron River over the years, compiling much of the information which appears in this guide. These inventory volunteers are Michael Appel, John Arkison, Vivienne Armentrout, Dea Armstrong, Barbara Bach, Pam Bailey, Nancy Jane Bailey, Jim Ballard, Jean Balliet, Tom Beauvais, Kevin Bell, Jamie Bender, Anne Benninghoff, Chris Berdoll, Cynthia Bida, Jeff Bieszki, Sally Bila, John Bingamon, Jan Blain, Dorothy Blanchard, Laura Blanchet, Jim Bonar, Don Botsford, Stephen Bowler, Bruce Bowman, Mark Brahce, Jim Breck, Lisa Brush, Jim Buckmaster, Russ Butler, David Cappaert, Carrie Cartwright, Trudy Casimir, Frank Casimir, Tim Cavnar, Mary Cavnar, Don Chalfant, Dee Chapell, Tom Chapell, Lathe Claflin, Kathy Claflin, Patrick Clancy, Ken Clark, Canny Clark, Jeff Clevenger, David Collopy, Daniel Comella, Mike Conboy, Joanne Constanides, Bill Cooke, Jacqueline Courteau, Suzanne Custons, Phil D'Anieri, Judy Dean, Jim Deigert, Bob Dennis, Ben Dickenson, Jim Dickinson, Judy Dluzen, Joan Doman, Cecily Donnelly, Betsy Dugan, Susan Ellis, Kevin Elmiger, Mary Fitts, Rick Foster, Irene Frentz, Harry Fried, Mark Gilbert, Mary Glass, Brian Glass, Karen Glennemeier, Bill Grams, Judy Gray, Dunrie Greiling, Kathy Guerreso, Don Hach, Natalie Harmon, Jillian Harris-Cowan, Will Harris-Cowan, Michael Hathaway, Ted Hejka, Connie Hertz, Kate Heywood, Steve Hiltner, Rob Hollenbeck, Sarah Horner, Trudy Hughes, Rebecca Hull, Jonathon Hull, Shirley Jankelevich, Jennifer Jaworski, Phil Jobson, Sharon Johnson, Laura Jones, Janet Kahan, Ethan Kane, Gail Karr, Sindi Keesan, Nopporn Kichanantha, Mike Kielb, Randy Knibbs, Abbey Koehn, Paul Kosnik, Anita Kraemer, Ruth Kraut, Maggie Kronk, Laura Krueger, Roger Kuhlman, Karen Land, Lisa Lava-Keller, Ann Lund, Dave Lyzenga, Megan Lyzenga, Jennifer Maigret, Adrienne Malley, Jeff Maple, Paul Marangelo, Andre Marcoux, JoAnn Marcoux, Maureen Martin, Bob Masta, Laurie McCall, Katherine McKee, Marilyn McKelvey, Paul McKelvey, Molly McReynolds, Kallie Michels, Bill

Minard, Sharon Minnick, Sally Moore, Elizabeth Moray, Jodi Mullet, Pauline Nagara, Melissa Nickles, Gwen Nystuen, John Nystuen, Eric Olson, Kris Olsson, Jeannine Palms, Janice Pappas, Mary Pappas, Larry Paris, Steve Parrish, Mercer Patriarche, Jerry Paulissen, Gerry Paulson, Sue Phare, Harvey Pohnert, Dave Pollock, Ralph Powell, Barbara Powell, Carl Presotto, Rob Pulcifer, Cynthia Radcliffe, Neil Richards, Bev Richards, Lindsey Richards, Joel Robbins, John Robbins, Renee Robbins, Joanna Rodgers, Norm Roller, Bob Rommel, Tracy Samilton, Paul Sanschagrin, Nancy Schiffler, Janet Schoendube, Barb Schumacher, Greg Sharp, Beverly Shepard, Nancy Shiffler, Kathleen Singer, Lucia Skoman, Frank Skoman, Cecille Smith, Jackie Smithers, Laura Spears, Rebecca Staffend, Taryn Stejskal, Pat Stejskal, Deb Stenkamp, Mike Stippanovich, Mary Stock, Nancy Stone, Allison Stupka, John Swales, Charles Swift, Stefan Szumko, Dan Thiry, Jean Thomson, Irwin Titunik, Amanda Todd, Terri Torkko, Monica Touesnard, Frances Trail, Clint Turbeville, Thelma Valenstein, Sarah Van Tiem, Rick Vetsch, Keistel Vetsch, Kim Waldo, Jessica Wallace, Don Ward, Camille Ward, Helen White, Ben Wieland, Leslie Williams, Ellen Wilson, Mary Wise, Devorah Wolf, Elizabeth Wood, Roger Wykes, Barbara Wykes, Sean Zera, Karen Zera, and Tony Zink. Likewise, we wish to thank the numerous volunteers who regularly attend our work days in these natural areas, helping to restore the native communities they contain. Unfortunately, we are unable to individually name all of these volunteers here, but it is their interest in, and support of, our natural areas that has made this guide possible.

The guide was written by Natural Area Preservation (NAP) staff of the City of Ann Arbor, Department of Parks and Recreation, namely: David Borneman, Malin Ely, Bridget Fahey, Tim Howard, Mike Kielb, David Mindell, Catriona Mortell, Deb Paxton, and Bev Walters. Malin Ely and Amie Ottinger were responsible for editing revisions to the guide. Illustrations were done by Mike Kielb, Greg Vaclavek (who also illustrated the cover), Susan Kielb, Malin Ely and Sheila Mortell. We received valuable input and editorial comments from many Parks and Recreation staff members. The writers also relied heavily on the work of other current and former NAP staff members who have been involved with related projects: Dea Armstrong, Michelle Barnwell, Trish Beckjord, Chris Berdoll, Joseph Bogaard, Kristie Brablec, David Cappaert, Kirstin Condict, Eric Crawford, Irene Frentz, Vicki Gollwitzer, Tom Hulleberg, Brian Killian, Fred Kraus, Katie LaCommare, Mike Levine, Tammie Merschat, Sally Moore, David Pollock, Gideon Porth, Chris Rickards, Cara Rockwell, Bill Schneider, Kathy Sorensen, Mallory Tackett, Greg Vaclavek, Stacy Vinson, Dave Warners, Robert Williams and Alan Wolf. Thanks also to NAP staff Courtney Babb and Jennifer Maigret for their final product assistance.

We would like to thank the present and former members of the Park Advisory Commission, and the rest of the Parks and Recreation Department, who worked to preserve these natural areas as park land. Finally, we give special thanks to the citizens of Ann Arbor who generously supported the Department's 1993-1998 and 1998-2003 Park Maintenance and Repair Millages, which provides all funding for the Natural Area Preservation Division.

TABLE OF CONTENTS

INTRODUCTION

This book has been written with the nature lover in mind. The casual hiker will find useful trail information and descriptions of the natural communities along the way. A more studious amateur naturalist will be able to use the plant, bird and butterfly lists to help in their studies. Families will be able to have a reference for their explorations of the natural world. In this section the Natural Area Preservation Division and the thirteen natural areas of the Huron River corridor are introduced. Also described is the concept of "natural communities" and an explanation of some of the terms used throughout this guide. The section concludes with a description of park rules and some of the safety issues to consider when visiting the natural areas of the Huron River corridor.

THE DEPARTMENT OF PARKS AND RECREATION

The Natural Area Preservation (NAP) Division of the City of Ann Arbor, Department of Parks and Recreation was formed in 1993, following ballot approval of the 1993–1998 Park Maintenance and Repair Millage. Community support continues for the program with the ballot approval of the 1998–2003 Park Maintenance and Repair Millage. The mission of NAP is to protect, restore, and champion Ann Arbor's natural areas, especially those in the City's parks system. As part of that mission, staff and volunteers have been working to inventory the plants, amphibians, butterflies, and breeding birds of undeveloped Ann Arbor. This information helps us to identify critical habitats and to more effectively manage and restore the many significant natural areas within the park system.

Wild cucumber (Echinocystis lobata) vines have distinctive spiny green fruits, like bladders, which float and can help to disperse the plant's seed along waterways.

NATURAL AREAS OF THE HURON RIVER CORRIDOR

The thirteen natural areas described in this guide include eleven city parks, one county park, and one university arboretum. Together they comprise over 630 acres of undeveloped green space along the banks of the Huron River, stretching from Foster Bridge on Maple Rd. on the northwest fringe of Ann Arbor, through the center of town and past US–23 to the southeast—a river distance of approximately nine miles. The city park system is currently changing the names of the city's undeveloped parks from "Park" to "Nature Area." In this guide we will use both Park and Nature Area when referring to a specific location owned by the City of Ann Arbor Department of Parks and Recreation.

Natural Areas of the Huron River Corridor in Ann Arbor

NATURAL AREA	OWNERSHIP	LAND ACREAGE
Argo Nature Area	City	22.5
Bandemer Nature Area	City	36.9
Barton Nature Area	City	83.6
Bird Hills Nature Area	City	147.7
Cedar Bend Nature Area	City	19.5
Foster (Barton Nature Area)	City	24.9
Furstenberg Nature Area	City	35.3
Gallup Park – Wet Prairie	City	10.1
Kuebler Langford Nature Area	City	31.0
Nichols Arboretum	City/University	142.2
Parker Mill/Forest Nature Area	City/County	43.4
Ruthven Nature Area	City	20.5
South Pond Nature Area	City	13.8
TOTAL ACREAGE =		**631.4**

These thirteen natural areas are strung out like pearls along either side of the Huron River (see map inside back cover), which is a state-designated "natural scenic river" just a short distance upstream to the northwest of Ann Arbor. Linking the natural areas are numerous additional parcels of public land owned by either the City of Ann Arbor or the University of Michigan. As a result, if you were to paddle a canoe from one end of this corridor to the other, you would always find public land on at least one bank of the river!

NATURAL COMMUNITIES

This nearly continuous band of open green space helps buffer the river from the impacts of urbanization and provides scenic recreational opportunities for the public. It also provides critical habitat for plants and wildlife and serves as a corridor for their movement. Although the green space is nearly continuous along the river, its character varies greatly. Some sites are wooded, others are open prairie remnants, still others are wetlands.

Throughout this guide, we identify these different types of green space as different "natural communities." A natural community is a grouping of plants and animals commonly found living together. The types of animals found on a site depend primarily on the vegetation and water present there. In turn, the existing plant species are determined by many non-living factors including soils, hydrology, slope, aspect, and disturbance. Because water plays such a key role in defining each natural community, its presence or absence is indicated in many of the names given to these communities, for example: wet prairie, mesic (moist) forest, or dry shrubland.

We hope that this guide will increase your enjoyment of, and appreciation for, the natural areas you visit along the Huron River.

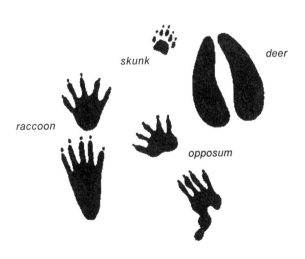

Many animal species live in the natural communities of the Huron River corridor. Tracks are most easily spotted in snow or soft mud.

HOW TO USE THIS GUIDE

This guide focuses on the natural communities you may encounter as you visit the various parks and other sites along the Huron River in Ann Arbor. It will help you locate the site, park your car or bicycle nearby, find the trailheads, and catch any seasonal natural highlights in the park.

The guide is organized with general overview information on the Huron River corridor first, followed by more detailed descriptions of the various natural communities in the corridor and the plants and animals which inhabit them. Individual park maps and descriptions follow near the end of the guide. A fold-out overview map of the entire Huron River corridor covered in this guide may be found inside the back cover. Consult this map to see the general location of each natural area in relation to the rest of the corridor.

Names

Throughout this guide, you'll find the common name of an individual plant species followed by its scientific name italicized in parentheses. The scientific name is usually composed of two words: the first is the genus name and is capitalized, the second denotes the individual species within that genus and is not capitalized. White oak (*Quercus alba*), black oak (*Quercus velutina*) and red oak (*Quercus rubra*) all share the genus name *Quercus*, but each has its own unique species name. Although common names may be easier for most people to understand, many species have several common names and some common names are shared by two or more plants. Therefore, in order to identify and communicate accurately about an individual species, the scientific name for the plant is always included. There are, however, instances when the individual species of oak is not important and all oak trees are being referenced. In that case, the genus name *Quercus* is used alone.

Just like plants, each animal species also has a unique two-word scientific name, but it is not used in this guide. Unlike plants, most animals—such as spring peepers, Red-winged Blackbirds, and Monarch butterflies—have a common name which is fairly standard in general use. Therefore, this common name alone will accurately communicate which species is being discussed.

Regarding capitalization of common names, we have tried to be as consistent as possible, but there are different standards among botanists and birders. Common plant names are never capitalized here unless they refer to a proper noun, such as American elm or Jack-in-the-pulpit. Ornithologists have developed their own rules regarding the capitalization of bird names, and we have adopted their widely accepted practices.

CITY PARK RULES

The natural areas within the Huron River corridor are fragile sites. To help protect them, and to help you and others enjoy the parks safely, please observe the following city park rules adopted by the Ann Arbor Parks and Recreation Department:

- Alcoholic beverages are not allowed.
- Hunting and/or trapping of animals is prohibited.
- Fires are allowed only in designated areas.
- Driving or parking motor vehicles in any area of a city park other than along roads and in parking areas designated for public use is prohibited.
- Overnight camping or overnight parking are not allowed.
- All city parks are closed between 12:00 midnight and 6:00 am.
- Motorized boats are not to be operated in excess of "slow, no wake" speed.
- Swimming is prohibited in all park ponds, lakes, rivers, and streams.
- All dogs and other pets must be leashed and under control at all times. The person responsible for the animal is also responsible for collecting and removing its feces immediately.
- Feeding and/or harassing ducks, geese, and other wildlife is prohibited.

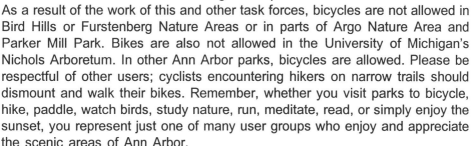

In 1994, a Mountain Bike Task Force was formed within the Department of Parks and Recreation to address the issue of mountain biking within parks. As a result of the work of this and other task forces, bicycles are not allowed in Bird Hills or Furstenberg Nature Areas or in parts of Argo Nature Area and Parker Mill Park. Bikes are also not allowed in the University of Michigan's Nichols Arboretum. In other Ann Arbor parks, bicycles are allowed. Please be respectful of other users; cyclists encountering hikers on narrow trails should dismount and walk their bikes. Remember, whether you visit parks to bicycle, hike, paddle, watch birds, study nature, run, meditate, read, or simply enjoy the sunset, you represent just one of many user groups who enjoy and appreciate the scenic areas of Ann Arbor.

Preserve natural features - enjoy flowers without picking them.

Act responsibly.

Respect other city park users.

Keep the parks clean & enjoyable for everyone - be responsible for your trash.

Stay on the established trails - many plants and animals are easily trampled!

Or, even more simply, "Parks belong to everyone; be a respectful guest."

For a complete listing of these and other city park rules, please contact the Department of Parks and Recreation office at 734-994-2780.

Please note that there are other natural areas in the Huron River corridor that are not city parks, but are owned or managed by the University of Michigan or by the Washtenaw County Parks and Recreation Commission. For more information on the rules in these natural areas, please contact the University of Michigan's Nichols Arboretum (734-998-7175), or the Washtenaw County Parks and Recreation Department (734-971-6337).

SAFETY CONCERNS

In addition to the park rules, we have several other suggestions for making your visit to the parks a safe one:

Poison ivy (*Toxicodendron radicans*) is found in many parks, and can be encountered in forests, old fields, or wetlands. Learn to identify its three leaflets in the spring, summer and fall, and its white fruits in the winter. When growing up a tree as a woody vine, it can also be identified by its fine hairlike rootlets which attach the vine to the tree trunk.

Probably the easiest way to avoid poison ivy is to STAY ON THE TRAILS. This will not only protect you, it will also protect sensitive mosses and wildflowers from being trampled, and keep the soil from getting compacted. If you are in a park with no designated trail system, you may be

Poison ivy (Toxicodendron radicans) leaves

able to follow a game trail, or simply pick your way slowly through the area, being careful about where you step.

Chiggers, bees, wasps and other stinging insects may also be encountered in the parks, but no more so than in any other outdoor setting. While it won't help with all of these "hazards," park users may want to prepare themselves with insect repellant before venturing into the natural areas.

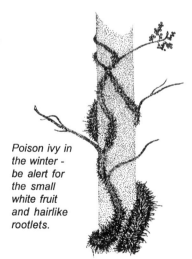

Poison ivy in the winter - be alert for the small white fruit and hairlike rootlets.

READING THE LANDSCAPE

If you were to step back and look at the landscape of Ann Arbor's Huron River corridor, you would see that it is a mosaic of hills, small valleys and flat plains. Each area along the corridor has slightly different conditions that determine the type of natural community that will be found there. The descriptions that follow will help you to read the landscape and understand its natural history. Since 1993, the Natural Area Preservation Division has also had an effect on the landscape through our restoration efforts in the natural areas. Thus, this section also includes a description of our restoration program and provides information about how to get involved in the natural areas of the Huron River area.

GLACIAL LANDFORMS

Southeast Michigan was shaped by the continental glaciers that covered the land more than 14,000 years ago. In Ann Arbor, you can see examples of at least three types of glacial deposits: moraines, kames, and outwash plains.

Moraines

Moraines are moderately–sloped hills which were formed at the edge of glacial ice. They are made up of glacial till, which is an unsorted mixture of gravel, sand, clay and silt carried in the glacier. These materials were scraped up by the glacier's forward movement, then deposited at the glacier's melting front. The melting ice can be thought of as a conveyor belt receding across the landscape. As long as the movement is constant, the glacial till is deposited in a uniform layer (called a ground moraine.) But if the glacier stalls at one point for a period of time, the glacial till is piled higher there, forming an end moraine. A series of end moraines may be formed when the melting or receding glacier stalls several times, forming concentric ridges.

Two end moraines are found in the river corridor in Ann Arbor: the Defiance Moraine (named for Defiance, Ohio, which marks its furthest extension) and the Fort Wayne Moraine (named for its reach to Fort Wayne, Indiana.) The lobe of the glacier that formed the Fort Wayne and Defiance Moraines moved in the general direction of northeast to southwest. Because Ann Arbor lies at what was the side of the glacial lobe rather than the front, the end moraines here are oriented northeast–to–southwest. Along the Huron River corridor in Ann Arbor, the best illustration of the Fort Wayne end moraine is the high ridge running through Bird Hills Park.

GLACIAL FORMATION IN THE ANN ARBOR REGION

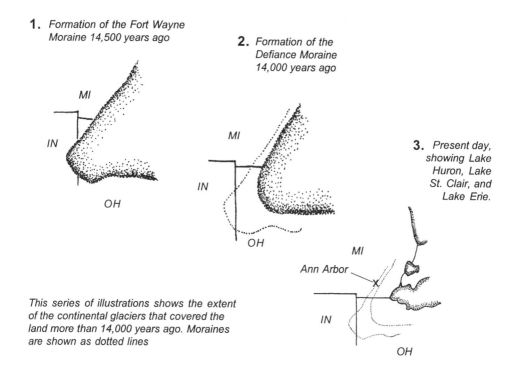

1. *Formation of the Fort Wayne Moraine 14,500 years ago*

2. *Formation of the Defiance Moraine 14,000 years ago*

3. *Present day, showing Lake Huron, Lake St. Clair, and Lake Erie.*

This series of illustrations shows the extent of the continental glaciers that covered the land more than 14,000 years ago. Moraines are shown as dotted lines

Kames

Kames were formed where there were holes, crevasses, or low spots in the glacial ice. As the ice melted, running water carried sediment into these openings. The heavier gravel and sand settled out at this point, while lighter silts and clays were carried off in the flowing water. When the ice surrounding these gravel and sand deposits finally melted away, the result was an inverse of the initial glacial topography: areas that had been lower were now be mounded with gravel and sand material sorted by the meltwater (forming kames,) while areas once occupied by glacial high points were now depressions. Without the support of the surrounding ice, the edges of the newly–formed kames slumped, forming steep–sided hills containing gravel and well–drained sand. The conical hill at Ruthven Nature Area is an example of a kame.

Outwash Plains

Huge rivers formed from the melting ice of the glaciers. The swiftly moving water carried sediment away from the glacial ice along river corridors. These rivers and streams changed course over time, braiding across the land in front of the glaciers, filling in low areas with sediment. In this way, river channels formed broad, flat outwash plains throughout the Ann Arbor area. Like kames, outwash plains contain gravel and sand sediments deposited by the flowing water. The rivers dropped these heavier materials first while carrying lighter clay and silt particles further downstream into glacial lakes and inland seas. The flat expanses of land at Furstenberg, Barton, and Bandemer Nature Areas and Gallup Park are all examples of outwash plains.

SOILS

The region's glaciers have sculpted two elements that are crucial in determining the composition of plants (and the animals they support) within our area: soils and the physical features of the landscape. These two factors interact to create habitat suitable only for particular plant communities. The ratio of sand to silt to clay, and the degree to which these particles have been sorted and layered within the soil profile dictate the availability of moisture and nutrients for plants that grow there. Sand, silt, and clay make up a continuum of decreasing soil particle size. Water rapidly drains through sandy soil because the large–sized sand particles have spaces between them. Sandy soils also tend to have a low nutrient content. At the other end of the spectrum, clay particles are so fine that there is little space for water movement between them. Thus, after rains or floods, water and nutrients tend to move downward very slowly, remaining available to plants for a long time. However, the oxygen available to the roots of plants growing in these heavy clay soils is often limited. The fine clay particles limit air pockets and make these soils particularly susceptible to compaction.

Big bluestem (Agropogon gerardii)

While moraines, kames and outwash plains are all examples of glacial deposition, they differ markedly in their soils. Kames and outwash features were deposited by running water, leaving heavier sand and gravel particles to form the sediment base. Moraines contain mixed sediment deposits because they were not sorted by flowing water. This basic difference has a large impact on the soil formations in the different natural areas of Ann Arbor. For example, the outwash plain at Barton Nature Area supports dry–tolerant old field and prairie communities on its sandy, well–drained soils, while the moraine at Bird Hills Nature Area just across the river supports moisture–loving plant communities on its rich soil.

PHYSICAL FEATURES

The glacial landform, the steepness and aspect of a slope, the position on that slope, and elevation are all key elements in determining the amount of moisture available to plants. Water runs off steep slopes much quicker than off gentle slopes or flat surfaces. Upper slope positions are typically drier than lower and bottom slopes, because water drains from the former to the latter. The slope's aspect also has a significant bearing on moisture: south–facing and west–facing slopes tend to be sunnier, hotter, and therefore drier than northern or eastern exposures. Elevation (ranging locally from 700–900 feet above sea level) can affect the temperature of the site, creating cold sinks or exposures to drafty conditions. For example, fog can often be seen collecting along the river near Barton Nature Area, at the foot of Bird Hills Nature Area. Cold air draining off of the morainal slopes contributes to the fog's formation.

RIVER LANDSCAPING

The defining feature of the Huron River corridor is, of course, the Huron River itself. As it winds its way through the city, it brings about subtle, daily changes in the landscape. In glacial times, however, the river was a tremendous force that scoured out the current river valley. During the formation of the Fort Wayne Moraine14,500 years ago, the river actually ran northwest (what would today be "upstream"), extending out into broad, low areas and carving into steeper, morainal banks.

Today, as the river glides along its bed, the movement of water gently etches new channels even as it fills in the old. Water flows more quickly along the outer edge of a river bend than along the inside. The additional energy from fast– flowing water causes cuts in river banks and carries eroded soils downstream. These particles drop from the water as it loses velocity along the inner banks or as the river enters a larger body of water like a pond. As a bend in the river grows, it gradually forms an **oxbow** — a characteristically broad, curving arc (similar in shape to part of the old–fashioned yoke worn by oxen, hence the name). In time, this curve may become so extreme that its beginning and end intersect, creating a shortcut in the river channel, and leaving a C–shaped "oxbow lake" behind (see illustration on facing page). There are no oxbow lakes along this section of the river, but if you examine the locations of wetter communities in the sections of Barton Nature Area near the oxbow bend in the river, you can easily pick out the depressions and ridges of old river channels.

Once a relatively common occurrence, floods deposited silts, clays, and rich, organic materials throughout the lowlands and floodplains along the river. Today, however, the river is carefully controlled by numerous dams that regulate the irregular or uneven flows of former times. Upstream from the dams are impoundments, or ponds, such as Barton and Argo Ponds. These are gradually growing shallower as more sediment falls out of the artificially–slowed water.

FORMATION OF AN OXBOW AND OXBOW LAKE

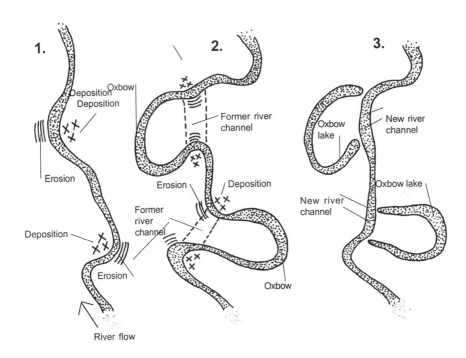

THE ROLE OF DISTURBANCE

Each of the natural areas described in this guide has been altered significantly by disturbances, both natural and human–caused.

Natural Disturbance

Natural disturbances are an integral part of our landscape. Historically, fire was a common feature of the Midwest landscape. Prairies and woodlands burned frequently as a result of lightning strikes. Prairies are perhaps the most fire–adapted ecosystem in our area. Fire controls the invasion of woody plants by stimulating prairie species to form a thick sod that prevents most shrub and tree seedlings from becoming established, while simultaneously killing the portion of the woody plant that is exposed to the fire. Because the roots of fire–adapted plants may grow twenty feet deep, these plants are not seriously harmed. Fire burns off dead litter (old grasses and leaves), allowing the sun to heat the soil quickly. Prairie plants tend to thrive in these warmer conditions. Woodland fires produced similar benefits to the native fire–adapted species.

Similarly, other natural occurrences have drastically altered the succession of plant communities and the landscape around us. The river corridor has been sculpted by floods, leaving new soil deposits, carrying patches of ground away, and dispersing seeds throughout the river floodplain. As the river ages, it meanders, leaving deposits in some spots while undercutting and eroding others. Wind storms and falling trees create openings in the forest canopy. The openings allow more light to stimulate the growth of some plants, again increasing the patchiness and diversity of habitats for plant and animal species within the woodlands.

Human Disturbance

Fire and Fire Suppression
The nature of human disturbances has changed considerably from historic times to the modern era. Fire was used extensively by Native Americans to clear the land for agricultural use, to flush animals from protected sites, or to create better habitat for game animals. While these practices altered the landscape, they largely served to perpetuate fire–dependent ecosystems. In contrast, modern day human disturbance has had a profoundly negative impact on fire–dependent ecosystems.

Igniting prescribed fires can help re–establish natural processes in fire–dependent ecosystems such as prairies.

Permanent dwellings built by European settlers in the early 1700s were incompatible with fire. People built houses in fire-dependent habitats and, in order to save their property, began to suppress fire, seeing it as a threat to their well–being. As cities grew, this practice became increasingly common. By extinguishing fires before they spread (in essence disturbing the natural disturbance patterns), we have allowed many fire–intolerant species to "out–compete" the prairie plants. Many of these fire–intolerant species are woody plants which form dense stands of shrubland where diverse, species–rich prairies once stood.

Dams
The damming of rivers altered the natural migration of watercourses and eliminated or modified historic flooding patterns. Consequently, plants that could tolerate sporadic high water have died off, while others have been flooded by the ponds and lakes created above the dams. Damming also affects fish populations by changing flow, temperature and habitat, creating conditions that allow troublesome zebra mussels to become established.

Non–Native Plants
Many aggressive, non–native plants have been introduced to our region, either inadvertently or as a by–product of promoting horticultural diversity. Many

species used in everyday landscaping have a devastating impact on native plants. People often purchase easy–to–grow plants for their yards, unaware that birds, wind, water, or gravity can carry the seeds into nearby natural sites where they may replace native plants.

There are no "untouched" natural sites within our public parks system. Past disturbance in one form or another has primed the landscape for colonization by invasive plants. Housing developments, road construction, and clearing for agricultural or timber production have dramatically increased the amount of "edge" in natural settings. It is through these edges that many of the invasive plant species gain a foothold. Once established along an edge, invasive plants begin to slowly migrate into the deeper, more pristine areas. These invasive plants typically have either a life history trait or a competitive edge in their new environment that gives them a growing advantage over the native species. As they expand their range, invasives may overwhelm the local plants that had evolved a dynamic balance of co–existence with the other local plants and animals, especially if natural processes like fire or flooding are no longer present to limit their invasion.

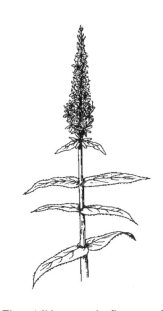

Purple loosestrife (*Lythrum salicaria*) and common buckthorn (*Rhamnus cathartica*) are two examples of this problem. Purple loosestrife, boasting a spectacular plume of purple flowers in late summer, was brought to the U.S. from Europe sometime before 1830. Since then it has been sold widely by landscape nurseries. The seeds began to spread far beyond their garden homes, creating thick colonies in wetland sites. This perennial plant grows so densely that it excludes other plants and makes formerly diverse areas unattractive to birds and other wildlife. Today it may be found along waterways and wet areas throughout the northern United States. While the sale of purple loosestrife has been prohibited in many states for several years (with the exception of "sterile" cultivars), its sale was not banned in Michigan until January of 1997. Things have moved a little faster in Ann Arbor, where in 1995 purple loosestrife was put on the city's invasive species list*, banning it from being planted in wetland mitigation sites.

The striking purple flowers of purple loosestrife (Lythrum salicaria) have made it a popular garden plant. Unfortunately, it has since escaped the cultivated sites, and has aggressively taken over stands of native vegetation, often to the exclusion of all other plants. In many places, including the state of Michigan, the sale of this de-structive plant has been banned.

*Ch.60 of the City Code - Wetland Preservation Ordinance

Common buckthorn (*Rhamnus cathartica*), like purple loosestrife, has a devastating impact on many native plant communities. This plant was sold horticulturally for windbreaks or as "specimen trees" (trees with interesting characteristics). In fact, the nation's largest common buckthorn was planted in Ann Arbor and still grows near Nichols Arboretum! Because buckthorn produces copious amounts of purplish–black fruit, leafs out early and keeps its leaves late into the fall, it was thought to be highly attractive throughout the year and was widely planted. Unfortunately, its longer leaf season and attractiveness to birds make buckthorn a keen competitor with native shrubs. Today, sadly, none of our natural sites is free of common buckthorn.

Non–Native Animals
It is not only invasive non–native plants that have spread into the diverse habitats of the Huron River corridor, but a wide variety of animals as well.

Butterflies
Among the butterflies we have two non–native additions to the North American scene: the Cabbage White and the European Skipper. The Cabbage White first appeared in Quebec, Canada around 1860 and rapidly spread across the continent. The European Skipper was accidentally introduced with nursery stock at London, Ontario, Canada, in 1910, and spread south, east, and west. Both species are now well established in the area and a common part of our butterfly fauna. A relative of the butterflies, the Gypsy Moth was purposefully introduced from Europe in Medford, Massachusetts, in 1869. It was hoped that a silk industry would follow. Unfortunately, this was not the case. Gypsy Moths have become established across eastern North America. Periodically, local populations increase dramatically and many forests are stripped of their leaves. While the idea of Gypsy Moth caterpillars consuming all of the leaves in a forest seems frightening, the moth outbreak usually lasts one to three years and most of the trees survive.

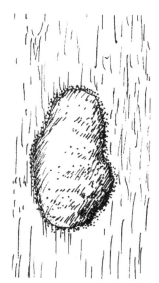

Look for 0.5"–1.5" yellow to orange to brown, powdery–textured gypsy moth egg cases on the bark of trees. These cases can contain 100–200 eggs apiece! Scraping them off into soapy water (not onto the ground where they can still hatch) can help destroy the eggs.

Fish

The common carp, a native of Asia, was introduced in 1877 by the United States Fish Commission in the hopes that it would become a plentiful food fish. It did become very plentiful, as it can tolerate a wide range of water quality, and will survive anywhere it can find food. Unfortunately, it never became a popular food item. The carp, a bottom feeder, causes much clouding of the water which many native species cannot tolerate. It also disrupts the nests of other fish species using the river bottom, and undercuts banks as a result of digging for food.

Zebra Mussels

The zebra mussel is a notorious invasive species which was accidentally introduced into the Great Lakes and is now frustrating water managers and boaters everywhere. It is a small, striped mussel about the size of a fingernail, which reproduces rapidly and blankets hard surfaces in the waters where it has been introduced.

Zebra mussels were first found in Ann Arbor's Huron River in 1994 where they were most likely introduced by hitching a ride on the boat of an unsuspecting recreational boater. Since then, they have been found in the Huron River in Barton, Argo, and Geddes Ponds (Gallup Park). The zebra mussel can be devastating to native species of clam, especially in areas directly downstream from impoundments (ponds created by the damming of streams). In addition to impacts on native clams and mussels, potential future impacts include an increase in water clarity which in turn results in increased weed growth as sun light penetrates to greater depths. In the Huron River, their numbers may be limited by the amount of hard substrate for them to attach.

Birds

After the domestic chicken, the Rock Dove, or pigeon, was the first non–native bird to be introduced into North America, arriving between 1606–1642. Today it is a common resident in cities, towns, and farmyards. In the middle of the 19th century the House Sparrow, a native of Europe, was introduced in Brooklyn, New York, and is now a common part of city birdlife across the continent. Among our field birds is the Ring–necked Pheasant, a successfully introduced game bird, recently bolstered by additions of the Sechuan Pheasant (the same species, but better adapted to the habitats of southern Michigan). Most recently, the House Finch has invaded Ann Arbor. Expanding west from Long Island, New York where birds where released in the late 1940s, the House Finch reached Ann Arbor in 1982 and is now one of the most common birds in the city.

Male House Sparrow

The starling is another introduced European species. Starlings were unsuccessfully introduced on a number of occasions between 1875 and 1890, but introductions in1890–1891 in Central Park, New York were successful and ultimately allowed the starling to become the most common city bird in North America. Among all introduced birds, the starling has perhaps had the greatest impact on our native birds. These highly aggressive cavity–nesting birds readily displace many native species. City parks once had bluebirds, wrens, flickers, and screech owls nesting where we now have only starlings.

Pets
Finally, while pets are not typically considered to be an invasive aspect of our natural settings, they do cause considerable damage to natural areas. Unleashed dogs have had a great impact on ground nesting birds, small mammals and fragile wildflowers in many urban settings. Feral and free–range cats cause even greater damage to local populations of small animals. Many cats regularly kill one or more small mammals and/or birds daily, greatly affecting populations locally. Obviously, responsible pet ownership is an important component to maintaining our natural environment.

NATURAL AREA PRESERVATION

The Natural Area Preservation (NAP) Division of Ann Arbor's Parks and Recreation Department works to restore the city's natural areas to their former splendor. While we are not able to accomplish this fully, we are working to reestablish corridors, such as the one described here along the Huron River, through which animals can move relatively unhindered, and plants can reproduce with minimal exposure to more aggressive weeds.

NAP staff, with the help of volunteers, also work to re–establish natural processes. Our prescribed ecological burns are an attempt to mimic the effects of natural fire, while conducted in a controlled, safe manner. In 1997, after only four years of our burn program, we have already begun to see a reduction of undesirable plants on burn sites.

Our program to remove invasive species is also a form of fire mimicry. By manually digging, cutting, and herbiciding unwanted plants from the region's natural areas, we are increasing the available light and decreasing the competition for local, native species of plants. Our goal is to continue expanding the regional diversity of native plants, and with them, native insects, birds, reptiles, amphibians, and mammals within our green spaces.

For more information about landscaping with native plants that will not invade our parks, contact the NAP office and ask about the native landscaping brochure series targeting specific native plants for landscaping in Southeastern Michigan.

HOW TO GET INVOLVED IN YOUR NATURAL AREAS

The local community can get involved with helping to make the natural areas of Ann Arbor healthier, cleaner spaces. Numerous volunteer opportunities exist with the Natural Area Preservation (NAP) Division of the Ann Arbor Parks and Recreation Department. NAP conducts volunteer workdays throughout the parks system two to three times a month from March through November. Volunteers remove invasive plants from the parks, collect seed to use in restoration work, or plant desirable, native vegetation in selected areas. We also encourage volunteers to work with us to implement broader management activities, often by "adopting" a park to work in or even helping to coordinate more comprehensive work activities there. In the spring, NAP conducts an intensive prescribed ecological burn program. We use carefully controlled fires in order to fulfill specific ecological objectives. These burns are done with the assistance of trained volunteers.

NAP also relies on volunteers to conduct yearly inventories within the parks. To date, we have inventoried populations of plants, frogs and toads, breeding birds, and butterflies in numerous sites throughout the city. We provide training in identification and inventory techniques, and help volunteers narrow their focus to a certain area within the park system. This is a wonderful opportunity to get to know the natural spaces within the Huron River corridor and beyond!

If you are interested in either inventory or stewardship activities, or if you have another idea for a park—or natural area—related project, feel free to call the Nature Area Preservation staff at 734-996-3266. Also use this number to reach our "stewardship hotline" and learn about upcoming volunteer events.

If you are interested in participating in other stewardship projects around our area, the following groups also have volunteer opportunities. Contact the volunteer coordinators at:

The Huron River Watershed Council	(734) 769-5123
Nichols Arboretum (University of Michigan)	(734) 998-7175
Michigan Chapter of The Nature Conservancy	(517) 332-1741
Parker Mill (Washtenaw County Parks)	(734) 971-6337

COMPONENTS OF THE NATURAL COMMUNITIES

This section describes the different natural communities of the Huron River corridor in detail, including descriptions to help identify each community and where examples of each may be located along the Huron River. The portion following the community descriptions describes plant species which may be spotted along the corridor at different times of the year. Descriptions of some of the resident animal species in the natural areas are included in the final portions of this section.

THE NATURAL COMMUNITIES

As you stroll through the natural areas along the Huron River corridor, you will encounter many different natural communities. These communities are classified here mainly by the kinds of plants they contain. In this guide, we have adapted a state–wide list of natural communities, developed by the Michigan Natural Features Inventory, for use in our local region. We did this by first recognizing that many of our small urban natural areas are less pristine than some of the larger natural areas around the state. Because of that we had to incorporate factors such as disturbance history into the community assessments. The definitions and descriptions on the following pages, therefore, are of use mainly in the Huron River corridor and may have limited use in other sites outside of this focus area.

Aside from heavy use areas and the open water of the river and other small ponds, this guide recognizes the following ten different community types in the Huron River corridor. This document uses two–letter abbreviations for the community types on the park maps in later sections of this guide.

EM — emergent marsh
WM — wet meadow
WP — wet prairie
DP — dry prairie
OF — old field
WS — wet shrubland
DS — dry shrubland
WF — wet forest
MF — mesic (moist) forest
DF — dry forest, including a variation sometimes referred to as "savanna"

With some exceptions, you are likely to find emergent marshes, wet meadows, wet prairies, wet shrublands, and wet or mesic (moist) forests when you are right along the river. A little further inland, old fields, dry prairies, dry shrublands, and mesic and dry forests become more common.

We should state that any attempt to categorize nature too strictly or draw boundary lines too precisely will quickly end in frustration. We can describe a "typical" mesic forest community, and find a site that closely matches that description. We can do the same for a dry forest community. But often as we try to draw an exact boundary line between those two communities, we realize that there is not a sharp border between the two, but rather a gradual transition. As you leave the center of the mesic forest community, patches of dry forest may begin showing up long before you actually reach what you feel is true dry forest. You may also find plants which are "supposed" to be found in dry forest actually growing quite nicely in mesic forest.

It is still helpful, however, to use these natural community designations in general terms. For example, there *are* general similarities between all the areas identified as mesic forest, and there *are* general differences between these forests and dry forests. But realize that natural communities exist along a continuum and nature is in a constant state of flux.

The table on the following page summarizes the location of these 10 natural communities in the natural areas of the Huron River corridor in Ann Arbor. The section that follows describes the natural communities in more detail, with notes on their common plant and animal species, the parks in which each community may be found, and other interesting tidbits about that community. The community descriptions are organized with more "open" types in the beginning (marsh, wet meadow, prairie, old field), followed by the more "closed" shrubland and forest community types.

In the plant lists accompanying the description of each natural community, species that are not native to southeast Michigan are noted with an asterisk (*). Also, you may want to refer back to page 4 to recall our use of scientific names.

NATURAL COMMUNITIES OF THE HURON RIVER CORRIDOR

	Argo	Bandemer	Barton	Bird Hills	Cedar Bend	Foster	Furstenberg	Gallup	Kuebler-Langford	Nichols Arboretum	Parker Mill	Ruthven	South Pond
Emergent Marsh (EM)	X		X	X		X	X				X		X
Wet Meadow (WM)		X	X			X	X			X	X		X
Wet Prairie (WP)								X					
Dry Prairie (DP)		X	X				X	X		X			
Old Field (OF)		X	X	X		X			X	X	X	X	
Wet Shrubland (WS)		X	X			X	X	X			X	X	X
Dry Shrubland (DS)											X		
Wet Forest (WF)		X		X		X	X			X	X		
Mesic (Moist) Forest (MF)	X	X	X	X		X	X	X	X	X	X	X	X
Dry Forest (DF)	X				X	X			X	X		X	X

EMERGENT MARSH

The emergent marsh community is found along pond, river and stream edges in shallow water. Emergent marshes are characterized chiefly by non-woody plants growing in year–round standing water (woody plants are trees, shrubs, and some vines; non–woody plants are all others). Emergent marshes contain many plants that, while rooted in the water, grow upright out of the water's surface. Such plants include narrow– and broad–leaved plants which often grow up to 6 feet tall. Marsh plants are especially adapted to soil that is saturated with water. Many other plants are unable to tolerate these saturated soils and do not grow in these areas.

A great diversity of dragonflies and damselflies is found in emergent marshes because of ample opportunities for feeding, breeding, and egg–laying. Additionally, large numbers of frogs and turtles are found here. Bird life is dominated by the Red-winged Blackbird and the Common Yellowthroat which are

The Common Yellowthroat is a typical summer resident of emergent marshes. Look for them nesting in the cat–tails (Typha).

common summer residents nesting within the cat–tails (*Typha*). It is here that you are most likely to find either the Sora or Virginia Rail, both nesting in small numbers. Great Blue Herons and Great Egrets are both commonly seen hunting for fish and frogs. In the spring and fall, Osprey are occasionally seen cruising overhead in search of fish. Among mammals, white-tailed deer frequently feed at the fringes of the marsh at sunrise. Muskrats are common as well, with their dens providing nesting mounds for Canada geese.

There are several ways to see emergent marshes within the corridor's natural areas. In Furstenberg Nature Area, an emergent marsh community can be viewed from the observation deck nearest the eastern end of the parking lot, and the footbridge to Barton Nature Area from Huron River Drive provides a good view of Barton's emergent marsh. Other emergent marshes may be viewed by canoe along the river. If traveling by boat, go carefully; the plants here are sensitive to the disturbance that motors and paddles may create.

Listed below are some of the plants that may be found in the Huron River corridor emergent marsh communities. Non–natives are noted with an asterisk (*).

Aquatic:
      buttercup (*Ranunculus*)
      common water meal (*Wolffia columbiana*)
      coontail (*Ceratophyllum demersum*)
      duckweed (*Lemna minor*)

duckweed (*Spirodela polyrhiza*)
Eurasian water-milfoil (*Myriophyllum spicatum*)*
pondweed (*Potamogeton*)

Grasses, Sedges, and Rushes:
bulrush (*Scirpus acutus, Scirpus atrovirens, Scirpus validus*)
fowl meadow grass (*Poa palustris*)
rice cut grass (*Leersia oryzoides*)
rush (*Juncus*)
sedge (*Carex lacustris, Carex comosa*)
spike rush (*Eleocharis*)

Wildflowers:
arrow-arum (*Peltandra virginiana*)
arrowhead (*Sagittaria latifolia*)
bur-reed (*Sparganium eurycarpum, Sparganium chlorocarpum*)
common cat-tail (*Typha latifolia*)
narrow-leaved cat-tail (*Typha angustifolia*)*
pickerel weed (*Pontederia cordata*)
purple loosestrife (*Lythrum salicaria*)*
water smartweed (*Polygonum amphibium*)
water plantain (*Alisma plantago-aquatica*)
waterlily (*Nymphaea odorata*)
yellow pond lily (*Nuphar variegata*)

*Emergent marshes are characterized by non–woody plants growing in year-round standing water. Note the arrow–shaped leaves of arrow– head (*Sagittaria latifolia*) in the right foreground.*

WET MEADOW

Found in low wet areas, wet meadows are open sites with few shrubs or trees. Standing water is common through the spring and early summer, but not year–round, even though the soil is always very moist. While grasses and tall wildflowers are common here, the dominant plants are sedges. Sedges are grass–like plants which can often be distinguished by their triangular stems ("sedges have edges").

Sedge–dominated natural communities provide abundant habitat for many species of butterflies, especially skippers and the Eyed Brown, with the Baltimore also found in areas where turtlehead (*Chelone glabra*) is growing. Hunting dragonflies are abundant, especially the larger Green Darner as well as other darner species, Eastern Pondhawk, and a variety of Meadowflies. An array of birds use the dominant shrubs— willows (*Salix*) and dogwoods (*Cornus*)— for nesting. The Yellow Warbler and Willow Flycatcher are most common among these, but you may also see Canada goose, Common Snipe, and American Woodcock. Common mammals include shrews, voles, and mice.

Shrews are common inhabitants of wet meadow communities.

Historically, people used wet meadows primarily for grazing livestock because the areas were too wet to be plowed for farming. The plants in wet meadows were also harvested as "marsh hay" which was used for livestock and for insulation in ice houses. These activities did not kill the roots of native plants, so in sites that have been abandoned, those plants have recovered and are now abundant. Seasonal flooding gives a competitive advantage to the moisture–tolerant plants found here. Periodic fires also maintain these sites by discouraging woody plants from invading the area.

To get a closer look at a wet meadow community, go to the Furstenberg Nature Area boardwalk trail, the Barton Nature Area oxbow area, or to Parker Mill.

Below are plants that may be found in wet meadow communities along the Huron River. Non–natives are noted with an asterisk (*).

<u>Shrubs</u>:
>	glossy buckthorn (*Rhamnus frangula*)*
>	red–osier dogwood (*Cornus stolonifera*)
>	silky dogwood (*Cornus amomum*)
>	willow (*Salix*)

<u>Grasses, Sedges, and Rushes</u>:
>blue joint grass (*Calamagrostis canadensis*)
>fowl manna grass (*Glyceria striata*)
>rush (*Juncus*)
>sedges (*Carex stricta, Carex pellita, Carex bebbii*)
>spike rush (*Eleocharis*)

<u>Wildflowers</u>:
>boneset (*Eupatorium perfoliatum*)
>Joe-pye weed (*Eupatorium maculatum*)
>marsh mint (*Mentha arvensis*)
>smooth swamp aster (*Aster firmus*)
>swamp-betony (*Pedicularis lanceolata*)
>tufted loosestrife (*Lysimachia thyrsiflora*)
>turtlehead (*Chelone glabra*)

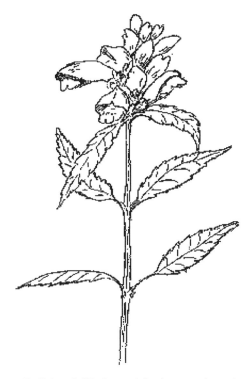

Turtlehead (Chelone glabra) grows in wet meadow communities among various types of sedges. Keep an eye out for the distinctive white flowers that look like turtle's heads.

PRAIRIES — WET AND DRY

Small pockets of prairies may be found throughout the Huron River corridor. These remnants are along the river and the railroad, representing the eastern reaches of the vast midwestern North American prairie system. Prairies are characterized by open areas of vegetation dominated by grasses. **Wet prairies** contain rich bottomland soils which are often inundated with water in the spring. Tall grasses, sedges and late summer–flowering plants characterize wet prairies. **Dry prairies** contain well–drained soils supporting tall grasses, late summer and fall–flowering plants, and sometimes a few scattered trees.

Adjacent to the river in Gallup Park is an area that supports wet–tolerant prairie plants. On higher sites along the railroad line, drier prairie remnants may be found. Prairie communities persist most vigorously where periodic fires have occurred. These fires help maintain fire–tolerant prairie plants while discouraging fire–sensitive woody plants from invading. This is one reason prairie remnants are found along railroads. In the past, sparks thrown by trains, especially steam engines, would ignite nearby dry grasses and start fires. The prairie communities in Barton, and Bandemer Nature Areas as well as part of Gallup Park are examples of these "railroad prairie remnants."

Prairies offer a wide array of plants for use by a great number of butterflies, both for feeding by adults and larvae, and for egg-laying. Coppers, Hairstreaks, and Blues are all found in prairie communities, especially if forested areas are nearby.

Solidago canadensis
Canada goldenrod

When snags (standing dead trees) are present, nesting opportunities for birds such as the Tree Swallow, Eastern Bluebird, House Wren, and Black–capped Chickadee abound. The illusive red fox may sometimes be seen in this community type.

Prairies are interesting to view year round. In spring, after a controlled fire, these areas rapidly recover and are blanketed with new green growth. While sedges are the most abundant plants in wet meadows, grasses dominate the prairies. Although some plants bloom in the summer, the spectacular season is the fall when prairie grasses are

Tall stalks of yellow–flowered Canada goldenrod (Solidago canadensis) grace the dry prairies around Ann Arbor in the fall.

tall and late–blooming wildflowers are in full color. The prairie grasses and flowers are remarkable for their striking display of colors, shapes and sizes. Throughout the winter, prairie plants lend color to the gray landscape, with varying hues of browns and golds. One way to tell native from non-native grasses is that many native prairie grasses change to a yellowish or reddish color in autumn while most non–native grasses either stay green or turn a less spectacular gray color.

Listed below are some of the plants that may be found in the Huron River corridor's prairie communities.

Wet Prairie: Visit Gallup Park for the only remaining example of this community in Ann Arbor. Non–natives are noted with an asterisk (*).

Shrubs and Vines:
       glossy buckthorn (*Rhamnus frangula*)*
       groundnut (*Apios americana*)
       hog peanut (*Amphicarpaea bracteata*)
       pussy willow (*Salix discolor*)

Grasses, Sedges, and Rushes:
       big bluestem (*Andropogon gerardii*)
       cordgrass (*Spartina pectinata*)
       fringed brome (*Bromus ciliatus*)
       sedge (*Carex*)
       wild rye (*Elymus virginicus*)

Wildflowers:
       angelica (*Angelica atropurpurea*)
       cowbane (*Oxypolis rigidior*)
       Culver's root (*Veronicastrum virginicum*)
       Missouri ironweed (*Vernonia missurica*)
       northern bedstraw (*Galium boreale*)
       prairie dock (*Silphium terebinthinaceum*)
       purple loosestrife (*Lythrum salicaria*)*
       purple meadow–rue (*Thalictrum dasycarpum*)
       starry false Solomon's seal (*Smilacina stellata*)
       swamp saxifrage (*Saxifraga pensylvanica*)
       tall sunflower (*Helianthus giganteus*)

Ferns:
       marsh fern (*Thelypteris palustris*)
       sensitive fern (*Onoclea sensibilis*)

Dry Prairie: Barton, Bandemer, and Furstenberg Nature Areas and Gallup Park have good examples of dry prairie communities. Non–natives are noted with an asterisk (*).

<u>Trees:</u>

black oak (*Quercus velutina*)
white oak (*Quercus alba*)

<u>Shrubs or Vines:</u>

hill-side blueberry (*Vaccinium pallidum*)
prairie willow (*Salix humulis*)

<u>Grasses, Sedges, and Rushes:</u>

big bluestem (*Andropogon gerardii*)
little bluestem (*Schizachyrium scoparium*)
switch grass (*Panicum virgatum*)
Indian grass (*Sorghastrum nutans*)

<u>Wildflowers:</u>

blazing star (*Liatris scariosa, Liatris aspera*)
butterfly–weed (*Asclepias tuberosa*)
hairy beard–tongue (*Penstemon hirsutus*)
showy goldenrod (*Solidago speciosa*)
smooth aster (*Aster laevis*)
spotted knapweed (*Centaurea maculosa*)*
sweet–clover (*Melilotus alba, Melilotus officinalis*)*
thimbleweed (*Anemone virginiana, Anemone cylindrica*)
yellow coneflower (*Ratibida pinnata*)

<u>Ferns:</u>

bracken fern (*Pteridium aquilinum*)

The dry prairie community of the Huron River corridor is an open area with well–drained soils supporting tall grasses, latesummer and fall–flowering plants, and sometimes a few scattered trees.

OLD FIELD

The old field community is typically a relatively open site with abundant wildflowers and grasses and scattered shrubs. Most old fields were once cleared and plowed as agricultural land or grazed heavily as pasture. After farmers abandoned these sites, both native and non–native plants re–colonized. Those plants whose seeds travel far in the wind (or the stomach of an animal) often arrive first while others take much longer to arrive. Some seeds may have persisted in the soil from before the disturbance. Hence old field sites are areas of transition and continue to change as more species arrive at the site and alter it with their presence. Depending on the disturbance pattern that befalls each site, some old fields may remain open for very long periods of time, eventually reverting back to prairie or meadow. For other old fields, the process of succession may continue with the influx of shrubs and trees until the site develops into a mature forest.

In old fields, butterflies such as the Common Sulphur and Orange Sulphur are sometimes abundant. Bee–balm (*Monarda fistulosa*), aster (*Aster*), ox–eyed daisy (*Chrysanthemum leucanthemum*), and sweet-clover (*Melilotus alba*) are all highly attractive to a variety of butterflies, offering nectaring opportunities throughout the summer and fall. The Black Swallowtail butterfly is often seen near Queen Anne's lace (*Daucus carota*) which it uses for egg–laying. The overgrown shrubby nature of old fields provide the greatest opportunity for the Brown Thrasher, a rather uncommon bird species in our area. You may also see Ring–necked Pheasant and Eastern Towhee in old fields.

Dipsacus

Teasel (Dipsacus) was first brought to the U.S. to help comb (or "tease") wool. Now it invades old fields.

To get a closer look at an old field community, go to Barton and Kuebler Langford Nature Areas, or Parker Mill Park.

Listed below are plants that may be found in the Huron River corridor old field communities. Non–natives are noted with an asterisk (*).

<u>Shrubs:</u>
      autumn olive (*Elaeagnus umbellata*)*
      gray dogwood (*Cornus foemina*)
      honeysuckle (*Lonicera*)*

Grasses, Sedges, and Rushes:
  Canada bluegrass (*Poa compressa*)*
  Kentucky bluegrass (*Poa pratensis*)*
  quack grass (*Agropyron repens*)*
  redtop (*Agrostis gigantea*)*
  tall fescue (*Festuca arundinacea*)*
  timothy (*Phleum pratense*)*

Wildflowers:
  aster (*Aster*)
  bee–balm (*Monarda fistulosa*)
  black medick (*Medicago lupulina*)*
  Canada goldenrod (*Solidago canadensis*)
  lawn prunella (*Prunella vulgaris*)*
  ox–eye daisy (*Chrysanthemum leucanthemum*)*
  Queen Anne's lace (*Daucus carota*)*
  white sweet–clover (*Melilotus alba*)*
  yellow sweet–clover (*Melilotus officinalis*)*
  hawkweed (*Hieracium*)*

*The nectar of bee–balm (*Monarda fistulosa*), also called wild bergamot, attracts butterflies to old field communities throughout the summer and fall.*

SHRUBLANDS— WET AND DRY

Two types of shrub communities are found in the Huron River corridor. **Wet shrublands**, often with standing water during part of the year, are adjacent to the Huron River, and to ponds, wet meadows and marshes. **Dry shrublands** are often found in areas that were once open; either old fields or dry prairies. Often thick with woody species, few other plants can get established under the dense shade of shrublands. In some areas, non–native shrubs such as buckthorn (*Rhamnus*) and honeysuckle (*Lonicera*) are invading native communities, especially when natural processes such as seasonal flooding or periodic fires have been absent for some time.

A wide variety of birds use willow (*Salix*) and dogwood (*Cornus*) shrubs for nesting. These include the stunning Indigo Bunting, Gray Catbird, Yellow Warbler, and Willow Flycatcher. Areas with hawthorn (*Crataegus*) often have a large number of nesting American Robins and Mourning Doves. The presence of large trees also offers nest sites for Eastern Kingbirds and Baltimore Orioles.

American robins are common in shrubland communities that contain hawthorns (Crataegus).

Listed below are some of the plants that may be found in the Huron River corridor shrubland communities.

Wet Shrubland: To get a closer look at a wet shrubland community, search out damp areas in Barton, Furstenberg or Bandemer Nature Areas or Parker Mill Park. Non-native species are noted with an asterisk (*).

Shrubs:
elderberry (*Sambucus canadensis*)
glossy buckthorn (*Rhamnus frangula*)*
guelder rose (*Viburnum opulus*)*
nannyberry (*Viburnum lentago*)
ninebark (*Physocarpus opulifolius*)
red–osier dogwood (*Cornus stolonifera*)
silky dogwood (*Cornus amomum*)
willow (*Salix*)

Wildflowers:
forget–me–not (*Myosotis scorpiodes*)*
fringed loosestrife (*Lysimachia ciliata*)
golden ragwort (*Senecio aureus*)
skunk–cabbage (*Symplocarpus foetidus*)

Dry Shrubland: To get a closer look at a dry shrubland community, go to upland areas of Parker Mill Park. Non–native species are indicated with an asterisk (*).

Shrubs:
>American hazelnut (*Corylus americana*)
>autumn olive (*Elaeagnus umbellata*)*
>common buckthorn (*Rhamnus cathartica*)*
>gray dogwood (*Cornus foemina*)
>hawthorn (*Crataegus*)
>honeysuckle (*Lonicera*)*
>smooth sumac (*Rhus glabra*)
>staghorn sumac (*Rhus typhina*)

Wildflowers:
>bee–balm (*Monarda fistulosa*)
>tall agrimony (*Agrimonia gryposepala*)
>thimbleweed (*Anemone virginiana*)
>white avens (*Geum canadense)*

*Dry shrubland communities are often found in areas that were once open —like old fields or dry prairies. The fuzzy, antler–like branches of staghorn sumac (*Rhus typhina*) seen in the right foreground are typical of this community type.*

FORESTS— WET, MESIC, AND DRY

Many forest types occur throughout the Huron River corridor. Based on the water–holding capacity of the soil, different tree species (like all other plant species) tend to grow in different locales. These groupings can be roughly divided into three different forest community types: wet, mesic, and dry forests.

Of the three, the **wet forest** is most easily recognized. Found in flat, poorly–drained bottomlands along streams and the river, a wet forest can be expected to flood in the spring. Wet forests often contain tall trees with a dense closed canopy and few shrubs beneath. The **mesic forest** is better drained, but is still most typical in moister sites as well as the richer glacial till soil of moraines. Mesic forests support communities characterized by a closed to semi–closed tree canopy with scattered shrubs in the understory and a rich display of wildflowers in the spring. The **dry forest** grows on sites with the best drainage, usually in areas with a layer of glacial outwash underlying the soil. It is often dominated by oaks (*Quercus*). In instances where these trees are so scattered that the canopy is fairly open, allowing a moderate amount of sunlight to reach the ground, the term "savanna" is used to describe this type of dry forest.

"Savannas" are dry forests with open canopies and are typically dominated by oaks (Quercus).

For many animals, the most important feature of forested communities is the presence of deadfall or standing dead trees, and the cavities they offer. As a result, cavity–nesting birds abound. Among birds, chickadees, titmice, wrens, owls, and woodpeckers are common forest residents. The forest also offers nesting opportunities for the Scarlet Tanager which nests in the canopy of taller trees, and the Red–eyed Vireo, which nests in the canopy of under–story trees. Also look for Turkey Vultures, Cooper's Hawks, Broad–winged Hawks, Acadian Flycatchers, Blue–gray Gnatcatchers, and Wood Thrushes in and around forest communities. Among mammals, squirrels, raccoons, and opossum are commonly found using tree cavities in forested communities.

Chickadees are common residents of forested communities. These curious, fearless birds often approach humans.

Most of the forests in our area are "second growth," which means they were once logged or farmed then allowed to grow back on their own. Evidence of furrows, fences and foundations can sometimes be found amidst the trees. Generally, plants which were originally found in these forests have regenerated and now create an approximation of the original forest community. An exception to this is the wet forest at Parker Mill/Forest Park. The ground cover vegetation and the apparent lack of old fence remnants suggest that the area along the river and south of the railroad tracks was probably neither logged intensively or plowed.

Listed below are plants that may be found in the Huron River corridor forest communities. To view most of the wildflowers, visit these areas in the spring, before the leaf canopy shades the forest floor.

Wet Forest: Visit Parker Mill/Forest Park or Nichols Arboretum for a good example of wet forest. Non–native species are noted with an asterisk (*).

Trees:
>	American elm (*Ulmus americana*)
>	black willow (*Salix nigra*)
>	black maple (*Acer nigrum*)
>	black ash (*Fraxinus nigra*)
>	box elder (*Acer negundo*)
>	cottonwood (*Populus deltoides*)
>	hornbeam (*Carpinus caroliniana*)
>	red maple (*Acer rubrum*)
>	red ash (*Fraxinus pennsylvanica*)
>	silver maple (*Acer saccharinum*)
>	swamp white oak (*Quercus bicolor*)

Shrubs:

bladdernut (*Staphylea trifolia*)
elderberry (*Sambucus canadensis*)
glossy buckthorn (*Rhamnus frangula*)*
red–osier dogwood (*Cornus stolonifera*)

Grasses, Sedges, and Rushes:

bulrush (*Scirpus atrovirens*)
fowl manna grass (*Glyceria striata*)
sedges (*Carex lupulina, Carex retrorsa*)

Wildlowers:

fringed loosestrife (*Lysimachia ciliata*)
great blue lobelia (*Lobelia siphilitica*)
skunk–cabbage (*Symplocarpus foetidus*)
smooth hedge nettle (*Stachys tenuifolia*)
touch–me–not (*Impatiens capensis, Impatiens pallida*)

Ferns:

ostrich fern (*Matteuccia struthiopteris*)

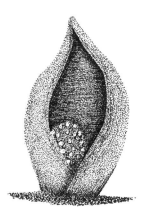

*Skunk cabbage (*Symplocarpus foetidus*) may be found growing on wet forest floors and is one of the first wild–flowers to emerge in the spring.*

Mesic Forest: Visit Bird Hills and Kuebler Langford Nature Areas for good examples of mesic forest.

Mesic (moist) forests are typically found in areas that have rich, moderately well–drained soils. They usually have a semi–closed canopy overhead, with some shrubs in the understory. "Spring ephemerals" carpet many mesic forests early in the season.

Trees:

       basswood (*Tilia americana*)
       black cherry (*Prunus serotina*)
       bur oak (*Quercus macrocarpa*)
       chokecherry (*Prunus virginiana*)
       hornbeam (*Carpinus caroliniana*)
       ironwood (*Ostrya virginiana*)
       red oak (*Quercus rubra*)
       sugar maple (*Acer saccharum*)
       white ash (*Fraxinus americana*)
       white oak (*Quercus alba*)

Shrubs and Vines:
>bristly greenbrier (*Smilax tamnoides*)
>gooseberry (*Ribes cynosbati*)
>gray dogwood (*Cornus foemina*)
>river bank grape (*Vitis riparia*)
>serviceberry (*Amelanchier*)
>wild black current (*Ribes americanum*)

Grasses, Sedges, and Rushes:
>long–awned wood grass (*Brachyelytrum erectum*)
>sedge (*Carex blanda*)
>silky wild rye (*Elymus villosus*)

Wildflowers:
>blue–stemmed goldenrod (*Solidago caesia*)
>cut–leaved toothwort (*Dentaria laciniata*)
>early meadow–rue (*Thalictrum dioicum*)
>false Solomon–seal (*Smilacina racemosa*)
>fragrant bedstraw (*Galium triflorum*)
>large flowered trillium (*Trillium grandiflorum*)
>wild geranium (*Geranium maculatum*)
>yellow trout lily (*Erythronium americanum*)

Ferns:
>lady fern (*Athyrium filix–femina*)
>maidenhair fern (*Adiantum pedatum*)
>spinulose woodfern (*Dryopteris carthusiana*)

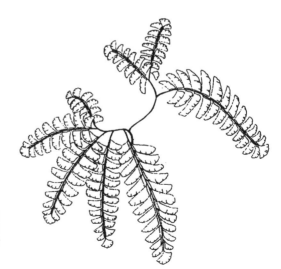

The lacy fronds of maidenhair fern (Adiantum pedatum) grace the goundcover layer of many mesic forest communities.

Dry Forest: Visit Foster, Cedar Bend, and South Pond Nature Areas to see examples of dry forest. Non-native species are indicated with an asterisk (*).

Trees:

      black oak (*Quercus velutina*)
      bur oak (*Quercus macrocarpa*)
      pignut hickory (*Carya glabra*)
      sassafras (*Sassafras albidum*)
      shagbark hickory (*Carya ovata*)
      white oak (*Quercus alba*)

Shrubs and Vines:

      downy arrow–wood (*Viburnum rafinesquianum*)
      maple–leaved viburnum (*Viburnum acerifolium*)
      hazelnut (*Corylus americana*)
      summer grape (*Vitis aestivalis*)
      witch–hazel (*Hamamelis virginiana*)

Grasses, Sedges, and Rushes:

      bottlebrush grass (*Hystrix patula*)
      Canada brome (*Bromus pubescens*)
      Pennsylvania sedge (*Carex pensylvanica*)

Wildflowers:

      bloodroot (*Sanguinaria canadensis*)
      early buttercup (*Ranunculus fascicularis*)
      early meadow–rue (*Thalictrum dioicum*)
      May apple (*Podophyllum peltatum*)
      nodding wild onion (*Allium cernuum*)
      round–lobed hepatica (*Hepatica americana*)
      shining bedstraw (*Galium concinnum*)
      wild ginger (*Asarum canadense*)
      wood anemone (*Anemone quinquefolia*)
      woodland sunflower (*Helianthus divaricatus*)

Ferns:

      bracken fern (*Pteridium aquilinum*)

PLANTS

Plant Inventories

The Natural Area Preservation staff, with the assistance of trained volunteers, has been conducting plant inventories of the Ann Arbor Parks since summer 1994. In the first three years, an impressive 978 species were found to be growing in Ann Arbor. The natural areas of the Huron River corridor boast an excellent array of these and offer a variety of new botanical adventures to the novice and seasoned naturalist alike. For whatever reason you venture forth, be certain to look for both the spectacular and the subtle displays found in our natural areas during the course of the year.

The plants mentioned below are just a few of the many that may be seen along the Huron River corridor in the various seasons of the year. The tables on pages 42–45 list selected plant species, their flower color, the communties in which they occur, and the parks in which you may find them. If you are interested in learning more about plants, see the *Recommended Reading* section at the end of the guide.

Spring

As you walk through the natural areas in early spring, the wildflowers are just beginning to show signs of life. First to peek through the melting snow in wetter spots is skunk–cabbage (*Symplocarpus foetidus*) with its unpleasant smelling flower stalk wrapped in a green and purple cloak at ground level. Swamp buttercup (*Ranunculus hispidus*) is all along the Huron River in the spring. Also in wet areas, the marsh marigold (*Caltha palustris*) makes quite a show with large golden flowers.

Jack–in–the–pulpit (Arisaema triphyllum*) is very distinctive in the sping. Look for its green and purple striped hood (called a* **spathe***) curving over an inverted, tonsil–like form called a* **spadix**.

The woodlands soon come alive with the delicate rue–anemone (*Anemonella thalictroides*), large–flowered trillium (*Trillium grandiflorum*) and Jack–in–the–pulpit (*Arisaema triphyllum*). Often dense carpets of yellow trout–lily (*Erythronium americanum*), pink spring beauty (*Claytonia virginica*), white toothwort (*Dentaria laciniata*) and pink wild geranium (*Geranium maculatum*) appear to light up the forest floor. The large umbrella–like leaves of the May apple (*Podophyllum peltatum*) and the lobed leaves of bloodroot (*Sanguinaria canadensis*) both unfold in April.

These early blooming wildflowers are called spring ephemerals (short–lived) because they emerge from the ground, flower before the overstory trees leaf out, then die back. This allows them to get more energy from the sunlight, storing it in underground bulbs. Before mid–June, most of the above–ground parts of these plants have withered. Not until the next spring will the underground storage organs again send up their leaves and flowers.

At some sites along the Huron River, the non–native dame's rocket (*Hesperis matronalis*), with large phlox–like flower clusters, has crowded out many native wildflowers. In more open, sunny areas, robin's plantain (*Erigeron pulchellus*) can be spectacular with its large blue–pink flowers. The redbud (*Cercis canadensis*) is a small tree that reaches its northern limit in the protected river valleys of southern Michigan. Its dark pink flowers can be seen along the banks of the Huron River in early spring.

Summer

As the seasons progress, the colorful display of spring blooming gives way to some of summer's more subtle flowers. Bottlebrush grass (*Hystrix patula*), appropriately named for its fruits arrayed in a bristly cylinder, and tick–trefoil (*Desmodium glutinosum*), which has a single stalk of small pink flowers above a cluster of bean–like leaves, both appear in drier forests.

In moist woodlands the green dragon (*Arisaema dracontium*) with large bizarre leaves and a long pointed flowering structure, looks as if it would be more at home in the tropics. The yellow flowers of golden ragwort (*Senecio aureus*) and, less frequently, the orange ones of Michigan lily (*Lilium michiganense*) can also be seen in these moist shady settings. Becoming all too common in wet open areas along the Huron River, the spires of the invasive purple loosestrife (*Lythrum salicaria*) signal the loss of native plant diversity.

Some plants are now more conspicuous in fruit than they were while in flower. Red baneberry (*Actaea rubra*) and Jack–in–the–pulpit (*Arisaema triphyllum*) both sport bright red berries, while white baneberry (*Actaea pachypoda*), also called doll's eyes, has white berries, each with a single black dot. On drier sunny

sites white–flowering spurge (*Euphorbia corollata*) and the brilliant orange flowers of butterfly–weed (*Asclepias tuberosa*) can often be seen together. Frequently found with them, and also along paths and in old fields, the lavender heads of bee–balm (*Monarda fistulosa*) emerge above leaves that have a pleasant minty fragrance when crushed.

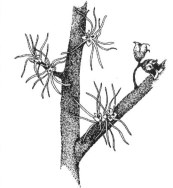

Fall

In the fall, yellow goldenrods (*Solidago*) and the white, blue and purple shades of asters (*Aster*) steal the show and can often be seen blooming into late October. This is a good time to explore some of the prairie areas to experience the colorful stands of tall prairie grasses like big bluestem (*Andropogon*

The delicate flowers of witch hazel (Hamamelis virginiana) often remain on the shrub's branches well into the winter.

gerardii) and Indian grass (*Sorghastrum nutans*). Also continuing to bloom at these sites, prairie dock (*Silphium terebinthinaceum*) has broad vertical leaves at the base of a tall yellow flower stalk that can exceed 6 feet.

In wetter areas the dark purple flowers of Missouri ironweed (*Vernonia missurica*) and the brilliant blue of the great blue lobelia (*Lobelia siphilitica*) illuminate the landscape. Some interesting forms occur in a few fall blossoms. The flowers of turtlehead (*Chelone glabra*) look very much like the head of that reptile in white, while the flowers of the closed, or bottle, gentian (*Gentiana andrewsii*) never open and resemble a cluster of small blue vials.

In the woodlands, wild lettuce (*Prenanthes*) with large arrowhead shaped leaves that were visible much earlier in the summer, is finally blooming with nodding white flower heads. Among the shrubs, the narrow yellow petals of the witch–hazel (*Hamamelis virginiana*) flower may persist even after the leaves have fallen. This is unusual because most woody plants flower in the spring.

Winter
Although winter signals the end of the flowering season, it is a great time to observe some of nature's less flamboyant treats. In wetter areas along the river, the bright twigs of the red–osier dogwood (*Cornus stolonifera*) contrast strikingly against the white snow.

Now you can more easily see the details of tree shape and bark without the distraction of leaves and flowers. Shagbark hickory (*Carya ovata*), with it's bark flaking away in long strips, can be spotted in most upland wooded areas. Look for the distinctive, pale green or white blotches on the bark of sycamore (*Platanus occidentalis*), growing in moist floodplain areas. Also watch for large old oaks (*Quercus*) with low growing branches. This form indicates that when the tree was young the area was quite open and sunny, even though now the site may be dominated by shade–grown trees whose trunks extend straight up to the forest canopy without branching. Note that the dense leaves of common buckthorn (*Rhamnus cathartica*) and honeysuckle (*Lonicera*), which were among the first to appear in the spring, continue to hang on until late November. Their fruits can be seen throughout the winter.

*Sycamore (*Platanus occidentalis*) bark*

Among the herbs, the flower heads of bee–balm (*Monarda fistulosa*) retain their minty fragrance throughout the winter and the fruits of dogbane (*Apocynum*) have dispersed their fluffy seeds leaving slender gracefully–opened pods. Another winter experience, though probably less appreciated, is finding the bristly fruits of stickseed (*Hackelia virginiana*), tick–trefoil (*Desmodium*) and enchanter's nightshade (*Circaea lutetiana*) clinging to pantlegs after an excursion outdoors.

Plants of the Huron River Corridor

COLOR KEY

- B = BLUE
- G = GREEN
- K = PINK
- L = LAVENDER
- O = ORANGE
- P = PURPLE
- Y = YELLOW
- W = WHITE

* NON-NATIVE SPECIES

Common name	Scientific name	Flower colors	Argo	Bandemer	Barton	Bird Hills	Cedar Bend	Foster	Furstenberg	Gallup	Kuebler Langford	Nichols Arboretum	Parker Mill	Ruthven	South Pond	Flowering period	Habitat
agrimony, tall	Agrimonia gryposepala	Y	x		x	x			x	x	x	x	x	x	x	Aug	OF, MF
alum root	Heuchera americana	G K	x								x	x	x	x			MF, DF
anemone, wood	Anemone quinquefolia	W	x	x		x			x	x	x	x	x		x	May–Jul	WF, MF, DF
aster, New England	Aster novae-angliae	P				x			x	x	x	x	x	x		Oct	OF, DS
avens, white	Geum canadense	W	x	x		x		x	x	x	x	x	x	x	x	Jun	MF, DF
baneberry, red	Actaea rubra	W		x		x	x		x		x	x		x		May–Jun	MF, DF
baneberry, white	Actaea pachypoda	W			x	x			x		x	x		x	x	May	MF, DF
bastard-toadflax	Comandra umbellata	W				x					x	x		x		May–Jun	DP, OF
bedstraw	Galium sp	W		x		x			x	x	x	x	x	x	x		A
bee-balm (bergamot)	Monarda fistulosa	L		x	x	x			x	x	x	x	x	x	x	Aug–Sep	DP, OF
big bluestem	Andropogon gerardii	G		x	x	x			x	x		x	x	x	x	Aug–Sep	WP, DP, OF
bindweed, hedge	Calystegia sepium	W K		x	x	x			x	x	x	x	x	x	x	Jun–Sep	OF, DS
blazing star	Liatris sp	P							x			x			x	Aug–Sep	DP, OF
bloodroot	Sanguinaria canadensis	W	x	x		x			x	x	x	x	x	x	x	Apr–May	DS, MF, DF
boneset, common	Eupatorium perfoliatum	W	x	x		x			x		x	x		x		Aug–Sep	WM, WP
bottlebrush grass	Hystrix patula	G	x						x		x	x		x		Jun–Aug	DS, MF, DF
bush-clover	Lespedeza capitata	W					x		x			x		x	x	Aug	DP, OF, DF
buttercup, early	Ranunculus fascicularis	Y							x		x	x				Apr–May	MF, DF
buttercup, swamp	Ranunculus hispidus	Y			x	x			x	x	x				x	May	EM, WS, WF
butterfly-weed	Asclepias tuberosa	O		x	x	x			x	x	x	x	x	x	x	Jul–Aug	DP, OF
coneflower, yellow	Ratibida pinnata	Y		x	x		x	x	x		x	x	x	x	x	Jul–Aug	DP
cow parsnip	Heracleum maximum	W	x		x	x		x	x	x				x	x	Jun	WM, MF
culver's root	Veronicastrum virginicum	W	x		x				x	x				x	x	Jul	WM, WP, DF
daisy, ox-eye*	Chrysanthemum leucanthemum	W	x	x	x	x		x	x	x	x	x	x	x	x	Jun–Aug	OF, DS

Plants of the Huron River Corridor

Common name	Scientific name	Flower colors	Argo	Bandemer	Barton	Bird Hills	Cedar Bend	Foster	Furstenberg	Gallup	Kuebler Langford	Nichols Arboretum	Parker Mill	Ruthven	South Pond	Flowering period	Habitat
dame's rocket*	Hesperis matronalis	B K	x	x	x	x	x	x	x	x	x	x	x		x		WF, MF, DF
Deptford pink*	Dianthus armeria	P										x		x		June–July	OF, DS, DP
dogbane	Apocynum sp	K W	x	x	x	x	x	x	x	x	x	x	x	x	x	June–July	DP, OF
dogwood, flowering	Cornus florida	W		x		x		x	x	x	x	x	x			May–June	MF, DF
dogwood, red-osier	Cornus stolonifera	W	x	x	x	x	x	x	x	x	x	x	x	x	x	June–Aug	WM, WS
elderberry	Sambucus canadensis	W	x	x	x	x		x	x	x	x	x	x	x	x	July–Aug	WS, WF
enchanter's nightshade	Circaea lutetiana	W	x	x	x	x	x	x	x	x	x	x	x		x	July–Aug	MF, DF
figwort, late	Scrophularia marilandica	P	x	x	x	x		x	x	x	x	x				July–Sept	MF, DF
flag, southern blue	Iris virginiana	B	x	x	x	x			x			x	x		x		EM, WM
flag, yellow*	Iris pseudacorus	Y	x	x	x	x		x	x	x	x	x			x		EM, WM
flowering spurge	Euphorbia corollata	W	x	x	x	x		x	x	x	x	x				July–Sept	DP, OF
forget-me-not*	Myosotis scorpioides	B	x	x	x	x			x	x	x	x		x	x	June–Sept	WS, WF
gentian, closed	Gentiana andrewsii	B	x	x				x	x	x	x	x			x	Aug–Oct	WM, OF
geranium, wild	Geranium maculatum	K	x	x	x			x		x	x	x	x		x	May–June	MF, DF
ginger, wild	Asarum canadense	P	x	x	x					x	x	x	x		x	May–June	MF, DF
golden Alexanders	Zizia aurea	Y	x	x	x			x		x	x	x	x		x	June–July	OF, WM, MF
golden ragwort	Senecio aureus	Y	x	x	x				x	x	x	x	x		x	May–Aug	WS, WF, MF
goldenrod	Solidago sp	Y	x	x	x	x	x	x	x	x	x	x	x	x	x	Aug–Oct	A
green dragon	Arisaema dracontium	G	x						x		x	x	x		x	June–July	WS, WF
hairy beard-tongue	Penstemon hirsutus	P W	x	x		x		x				x	x	x	x	June–July	DP, OF
hedge nettle, smooth	Stachys tenuifolia	K	x	x	x			x		x	x	x	x		x	July–Aug	WP, WF
hepatica	Hepatica sp	B K	x								x	x	x		x	April–May	MF, DF
horse-gentian	Triosteum aurantiacum	P		x	x							x	x		x	May–June	MF, DF
Indian grass	Sorghastrum nutans	G		x					x		x	x			x	Aug–Oct	DP
Indian pipe	Monotropa uniflora	W	x	x								x	x		x	July–Aug	MF, WP
ironweed, Missouri	Vernonia missurica	P		x	x				x		x	x	x	x	x	July–Sept	WM, WP
Jack-in-the-pulpit	Arisaema triphyllum	G P	x	x	x			x	x	x	x	x	x		x	May–June	WF, MF

Plants of the Huron River Corridor

COLOR KEY

- B = BLUE
- G = GREEN
- K = PINK
- L = LAVENDER
- O = ORANGE
- P = PURPLE
- Y = YELLOW
- W = WHITE

* NON-NATIVE SPECIES

** BOTH NON-NATIVE AND NATIVE SPECIES

Common name	Scientific name	Flower colors	Argo	Bandemer	Barton	Bird Hills	Cedar Bend	Foster	Furstenberg	Gallup	Kuebler-Langford	Nichols Arboretum	Parker Mill	Ruthven	South Pond	Flowering period	Habitat
Joe-pye weed	Eupatorium sp	K L	x	x	x	x	x	x	x	x	x	x	x	x	x	Aug–Sep	WM, WP
knapweed, spotted*	Centaurea maculosa	K P	x	x	x	x	x	x	x	x		x	x	x	x	May–Oct	DP, OF
lettuce, white	Prenanthes sp	W	x		x	x		x	x	x	x	x	x			Aug–Oct	MF, DF
lily, Michigan	Lilium michiganense	O	x		x	x			x	x		x	x	x	x	July	WS, WF
lobelia, great blue	Lobelia siphilitica	B	x	x	x	x			x	x		x	x	x	x	Aug–Sep	WP, WF
loosestrife, fringed	Lysimachia ciliata	Y	x	x	x	x	x	x	x	x		x	x	x	x	July–Aug	WM, WS
loosestrife, purple*	Lythrum salicaria	P	x	x	x	x	x	x	x	x		x	x	x	x	July–Sep	EM, WM
marsh marigold	Caltha palustris	Y	x		x	x	x	x	x	x	x	x	x	x	x	April–May	EM, WS, WF
marsh pea	Lathyrus palustris	P	x		x	x	x	x	x	x	x	x		x	x	May–July	WM, WP
May apple	Podophyllum peltatum	W	x		x	x	x	x	x	x	x		x			May	MF, DF
meadow-rue	Thalictrum sp	G Y	x		x	x	x	x	x	x			x	x	x	May–July	MF, DF
milkweed, swamp	Asclepias incarnata	K	x		x	x	x	x	x	x			x	x	x	July–Aug	EM, WM
monkey-flower	Mimulus ringens	B	x											x		Aug	EM, WM
mountain mint	Pycnanthemum virginianum	W						x			x	x	x	x	x	Aug	EM, WM
onion, nodding wild	Allium cernuum	K W	x								x	x	x	x	x	July	DP, DF
ostrich-fern	Matteuccia struthiopteris	G	x	x			x	x									WM, WF
pickerel weed	Pontederia cordata	P							x				x			July–Aug	EM
pokeweed	Phytolacca americana	G P		x	x	x			x			x					WM, MF
prairie dock	Silphium terebinthinaceum	Y										x				Aug–Sep	WP, DP
redbud	Cercis canadensis	K	x		x	x	x	x	x	x	x					April–May	MF, DF
richweed	Collinsonia canadense	Y		x	x	x		x	x		x	x		x			WF, MF
robin's plantain	Erigeron pulchellus	B K	x	x											x	May–June	DP, OF
rose, pasture	Rosa caroliniana	K	x		x	x									x	June	DP, OF
rue-anemone	Anemonella thalictroides	K W				x			x		x					April–May	DF, MF

Plants of the Huron River Corridor

Habitat key:

Code	Habitat
EM	= EMERGENT MARSH
WM	= WET MEADOW
DP	= DRY PRAIRIE
OF	= OLD FIELD
DS	= DRY SHRUBLAND
WS	= WET SHRUBLAND
DF	= DRY FOREST
MF	= MESIC FOREST
WF	= WET FOREST
A	= ALMOST ALL

Common name	Scientific name	Flower colors	Argo	Bandemer	Barton	Bird Hills	Cedar Bend	Foster	Furstenberg	Gallup	Kuebler-Langford	Nichols Arboretum	Parker Mill	Ruthven	South Pond	Flowering period	Habitat
skullcap, common	Scutellaria galericulata	B		x	x			x	x	x	x				x	June–Sept	WM, WP
skunk-cabbage	Symplocarpus foetidua	G P	x	x	x			x	x	x		x	x		x	April	WM, WF
smartweed, water	Polygonum amphibium	K	x	x	x			x	x	x						Aug–Sept	WM
snakeroot	Sanicula sp	G W	x		x	x		x	x	x	x	x	x			June–July	MF, DF
snakeroot, white	Eupatorium rugosum	W	x	x	x	x	x	x	x	x	x	x	x			Aug–Sept	WF, MF
Solomon's-seal	Polygonatum biflorum	W	x	x				x	x	x		x	x			May–June	MF, DF
Solomon's-seal, false	Smilacina racemosa	W	x	x	x	x		x	x	x	x	x	x		x	May–June	WF, MF, DF
spring-beauty	Claytonia virginica	K W						x	x	x	x	x			x	April–May	MF, DF
stickseed	Hackelia virginiana	B W	x	x	x	x	x	x	x	x	x	x	x			July–Aug	MF
sunflower	Heliopsis, Helianthus	Y	x	x				x	x	x			x		x	July–Sept	A
swamp-betony	Pedicularis lanceolata	Y							x						x	Aug–Sept	WM, WP
sweet-cicely	Osmorhiza sp	W	x	x	x		x	x	x	x	x	x	x		x	May–June	MF, DF
thimbleweed	Anemone sp	G W	x	x	x		x	x	x	x	x	x	x	x	x	June–July	DP, OF, DF
thistle**	Circium sp	K L	x	x	x		x	x	x	x	x	x	x		x	July–Sept	DP, OF
tick-trefoil	Desmodium glutinosa	K L	x	x	x		x	x	x	x	x	x	x		x	July–Aug	MF, DF
toadflax (butter & eggs)	Linaria vulgaris	Y						x			x	x	x		x	Sept–Oct	OF
toothwort	Dentaria laciniata	L W						x		x	x	x	x			April–May	MF, DF
touch-me-not	Impatiens sp	O	x	x	x			x	x	x	x	x	x		x	July–Sept	WF, MF, WM
trillium, common	Trillium grandiflorum	W								x		x				April–May	MF
trout-lily	Erythronium sp	Y						x		x	x	x	x		x	April–May	MF, DF
turtlehead	Chelone glabra	W		x				x	x	x		x	x		x	Aug–Sept	WM, WP
vervain, blue	Verbena hastata	B P	x	x	x		x	x	x	x	x	x	x		x	July–Sept	WM, WP
vervain, white	Verbena urticifolia	W	x	x	x			x	x	x	x	x	x		x	July–Aug	MF
violet	Viola sp	B P	x	x	x			x	x				x	x	x	April–June	WF, MF, DF
waterleaf	Hydrophyllum sp	P W						x	x			x	x			June–July	MF, DF
water-lily	Nymphaea odorata	W														June–Aug	EM
wood sage	Teucrium canadense	K L						x	x	x		x	x	x	x	June–Aug	MF

ANIMALS

Where water and land meet, there is bound to be a broad array of animals, both vertebrates and invertebrates. The varied terrestrial and aquatic habitats found along the Huron River provide resources for a wealth of fauna too numerous to cover completely in this guide. We have chosen a diverse mix of species you are likely to encounter during your visits to our natural areas. Damselflies, dragonflies, butterflies, fish, amphibians, reptiles, mammals, and birds are briefly covered here to give the amateur naturalist a place to start. For those wanting more information, we have included a *Recommended Reading* section at the end of the guide.

Snapping turtles can be spotted by looking for their triangular–shaped heads peering out of murky water along the Huron River and in large ponds.

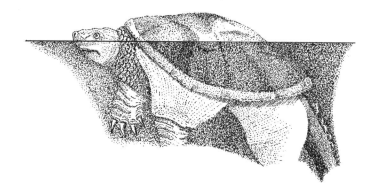

Damselflies and Dragonflies

The Huron River, its tributaries, and associated wetlands offer excellent habitat for a wide variety of damselflies and dragonflies. The two groups can easily be distinguished. All dragonflies and damselflies have four wings. The front wings, closer to the head, are the forewings; behind these are the hindwings. In the damselflies both the forewings and hindwings are shaped similarly, and are typically held closed above the back when perched. Damselflies are also much more lightly built than their close relatives the dragonflies. The forewings and hindwings of dragonflies are shaped differently and held straight out to the sides when the insect lands on something. There are at least 35 species of dragonflies found along Ann Arbor's Huron River corridor, in contrast to about 20 species of damselflies. Compared to the damselflies, dragonflies are strong fliers and much more robust in both the thorax and abdomen. Both groups are found in upland fields and marshes, and along small streams as well as the Huron River. All species are voracious predators and consume large numbers of smaller insects. Daily, a dragonfly can eat its weight in mosquitoes. When trying to identify these beautiful insects, observe three characteristics:

- wings— position and shape
- thorax (body)— color and markings
- abdomen (tail)— color and markings

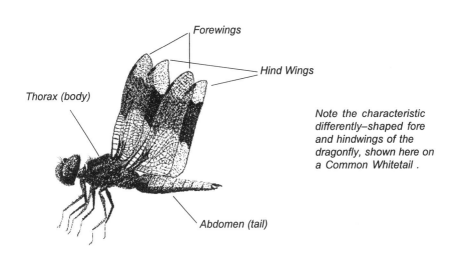

Note the characteristic differently–shaped fore and hindwings of the dragonfly, shown here on a Common Whitetail .

Damselflies

One of the most easily recognized damselflies is the Ebony Jewelwing, a species preferring streams and rivers and rarely seen around ponds or lakes. The male Jewelwing, our largest damselfly, has a metallic green abdomen and black wings. The female has a brown abdomen and wings which are cloudy gray with a distinct white dot near the tip of each forewing. This dot is called the stigmata, and is present in all damselflies and dragonflies, although in other species it is not nearly as distinct.

The bluets are another group of damselflies that are easily recognized. These very small, slender creatures have bright, sky–blue abdomens, marked with a variety of black dots and/or dashes. Most of the bluets require very close examination with a hand lens in order to determine the species.

Two other species easily identified are the Common Forktail and Violet Dancer. The Common Forktail's thorax and abdomen are yellowish–green with long black dashes above. At the tip of its abdomen there is a bright aqua dot. The Violet Dancer lives up to its name with an overall color of deep violet. Once again, as with most other damselflies, there are black markings along the abdomen.

Dragonflies

One of the largest and most easily identified dragonflies is the Green Darner. The thorax and head are bright lime–green and the abdomen is aqua with small black markings along the top. This species preys upon any insect smaller than itself, including other dragonflies and damselflies.

Two medium–large species with bluish–white abdomens occur in the area. The Twelve–spot is the larger of the two and has a wonderful array of markings on the wings. There are three black marks on each wing (hence the name twelve-spot) with three white marks on the forewing and two white marks on the hindwing. The color of the abdomen tends to be more blue than in the next common species, the Whitetail, whose abdomen is distinctly white with a blue cast. The Whitetail's wings, however, are quite different from the Twelve–spot. Each wing has a dark–brown dash extending out from the thorax, and a broad dark band extending through the center.

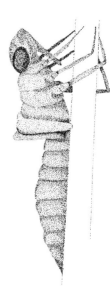

The smallest dragonflies in the area are the meadowflies. These are the bright red and yellow dragonflies found in large numbers in all of the meadows in close proximity to the river. As with the bluets, the meadowflies are difficult to identify to species. With practice, however, you should be able to identify the White–faced Meadowfly (with an obvious white face), Yellow–legged Meadowfly (reddish face and yellow legs), and the Ruby Meadowfly (yellowish face and black legs).

*The dragonfly **exuvia** is the shed outer skin, left behind as the insect increases in size.*

Finally, a large species of late summer and fall is the Black–mantled Glider. This is a very dark species, with mostly clear wings, except for a large dark area at the base of each hindwing (thus the alternate name, Black Saddlebags).

Butterflies

No group of animals is more strongly dependent on healthy and varied habitats than the butterflies. Not only do butterflies need a diversity of species from which to feed as adults, they also require specific **larval host plants**. These are the plants that adult butterflies lay eggs on and which ultimately become the food for the developing caterpillar. For instance, while a Monarch will feed on the nectar of a wide variety of flowers, a female will only lay eggs on milkweed (*Asclepias*). Because butterflies have such complex natural histories, many species will occur only in areas where specific habitat requirements are met *and* specific plants are present. Ann Arbor is fortunate to have may parks which meet both of these needs.

The life–history of a butterfly is quite interesting. From the egg emerges a very small caterpiller. As the caterpiller feeds, the skin is shed each time the caterpiller grows. The period between each successive shedding of skin is called an **instar**. Finally, the caterpiller reaches the final growth–stage and forms a **chrysalis**. This is similar to the cocoon of a moth and is where metamorphosis will occur. **Metamorphosis** is the process where the complete physical makeup of the caterpiller changes. After some time, from weeks to months, the adult butterfly emerges and the whole cycle begins anew.

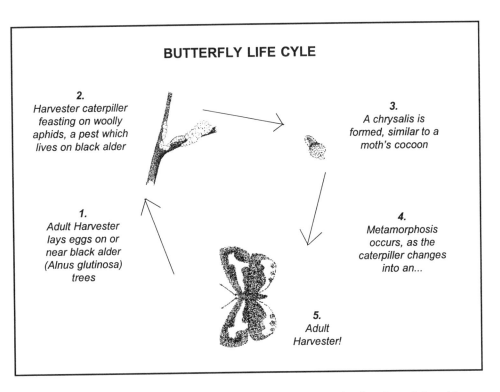

BUTTERFLY LIFE CYLE

2.
Harvester caterpiller feasting on woolly aphids, a pest which lives on black alder

3.
A chrysalis is formed, similar to a moth's cocoon

1.
Adult Harvester lays eggs on or near black alder (Alnus glutinosa) trees

4.
Metamorphosis occurs, as the caterpiller changes into an...

5.
Adult Harvester!

Two of the rarest butterflies in the state can be found in the Ann Arbor Huron River corridor. Interestingly, both of these species are dependent on non–native plants. The special concern Wild Indigo Duskywing is found in a number of parks and other areas nearby in the late summer and early fall. At one time, this butterfly was dependent on wild indigo (*Baptisia tinctoria*) as its larval host. As prairies gave way to development and wild indigo plants started disappearing, Wild Indigo Duskywings started disappearing too. But something interesting happened as freeways spread across the eastern United States—the Wild Indigo Duskywing prospered. The secret to the population boom was neither the access to freeway–travel, nor the accompanying development, but rather the introduced plant, crown vetch (*Coronilla varia*). Crown vetch was planted widely along banks to help stabilize the soil and prevent erosion, and this plant, which is in the same family as wild indigo, provided a new larval host for the butterfly. If you learn to identify crown vetch (which is, unfortunately, a highly invasive species) you will have an opportunity to find these unusual butterflies.

Another unusual butterfly dependent on an introduced plant is the Harvester. The plant in this case is the black or European alder (*Alnus glutinosa*) tree found growing along the banks of the Huron River. What is most interesting, however, is that the Harvester caterpillars do not feed on the plant, but harvest another species that is a pest to these trees— the woolly aphid. Woolly aphids appear as white fuzzy patches on the sides of the dark bark of the trees. The caterpiller of the Harvester is the only predatory butterfly caterpillar in Michigan. The adult Harvester will usually be found close to areas that have woolly aphids. When you find the aphids, start looking for the small orange and black butterfly.

More information on the habitat, larval host plant, and flight period of the common butterflies found along the Huron River is contained in the table on the following page.

The Buckeye is distinctive for the two, large black spots on its forewings which resemble eyes. This butterfly prefers open, sunny areas such as old fields and prairies.

COMMON BUTTERFLIES OF THE HURON RIVER AND ADJACENT LANDS

	Habitat	Larval Host Plant	Mar	Apr	May	June	July	Aug	Sep
Black Swallowtail	fields, secondary growth	parsley family			■	■	■	■	
Tiger Swallowtail	edges of woods	black cherry		■	■	■	■		
Spicebush Swallowtail	edges of woods	sassafras, spicebush			■	■	■	■	
Cabbage White	fields, open areas	crucifers		■	■	■	■	■	■
Common Sulphur	fields, open areas	white clover			■	■	■	■	■
Orange Sulphur	fields, open areas	pea family			■	■	■	■	■
Harvester	wooded wetlands	woolly aphids**				■	■	■	■
American Copper	fields, open areas	docks			■	■	■	■	■
Bronze Copper	wet meadows/ marshes	water dock, curled dock				■	■		
Banded Hairstreak	fields near woods	oaks, hickories					■		
Striped Hairstreak	fields near woods	cherry					■	■	
Eastern Tailed Blue	fields, open areas	pea family			■	■	■	■	■
Common Blue	fields, open areas	many hosts			■	■	■	■	
Great Spangled Fritillary	fields, open areas	violets				■	■	■	
Silvery Checkerspot	marshy wetlands	sunflowers, composites				■			
Pearl Crescent	fields, open areas	asters			■	■	■	■	■
Baltimore	marshy wetlands	turtlehead, plantain				■	■		
Question Mark*	woods	nettles, elm	■	■	■	■	■	■	■
Eastern Comma*	woods	nettles, elm	■	■	■	■	■	■	■
Mourning Cloak*	woods	willows, many others	■	■	■	■	■	■	■
Milbert's Tortoishell*	fields, open areas	nettles	■	■	■	■	■	■	■
Red Admiral*	fields, open areas	nettles		■	■	■	■	■	■
American Painted Lady*	fields, open areas	pearly everlasting			■	■	■	■	■
Red-spotted Purple	wet woods	cherry				■	■	■	
Viceroy	fields, wetlands	willows				■	■	■	■
Northern Pearly Eye	edges of woods	grasses				■	■	■	
Eyed Brown	marshy wetlands	sedges				■	■		
Appalachian Brown	wet woods	sedges					■		
Little Wood Satyr	fields near woods	grasses			■	■			
Common Wood Nymph	shrubby fields	grasses				■	■	■	
Monarch	fields, open areas	milkweeds			■	■	■	■	■
Silver-spotted Skipper	fields. open areas	black locust			■	■	■	■	■
Dreamy Duskywing	open woods	willows, poplars			■	■			
Sleepy Duskywing	open woods	oaks			■				
Juvenal's Duskywing	fields near woods	oaks			■				
Horace's Duskywing	fields near woods	oaks?			■			■	
Wild Indigo Duskywing	fields with crowned vetch	crowned vetch					■	■	
Least Skipper	marshes, wet meadows	grasses			■	■	■	■	
European Skipper	fields, open areas	timothy					■		
Peck's Skipper	fields, open areas	grasses				■	■	■	
Tawny-edged Skipper	fields, open areas	grasses				■	■	■	■
Little Glassywing	shrubby fields	purple top grass					■	■	
Mulberry Wing	marshes, wet meadows	sedges				■	■		
Hobomok Skipper	fields near woods	grasses				■	■		
Dun Skipper	marshes, wet meadows	sedges					■	■	

*These species overwinter as adults in crevaces and may be seen as early as January on very warm (60s) and sunny days.

**The caterpiller of the Harvester is the only North American carniverous species, thus lacking a host plant.

Fish

Fish are another type of wildlife found in the Huron River. They inhabit not only the river, but also many smaller streams and ponds. The Natural Area Preservation Division has not conducted any inventories of the fish life in the Huron River. Still, because interest in fish and fishing is high, we have included some information on local fish populations gathered by the Michigan Department of Natural Resources.

The species of fish listed on the next page are most likely to be found in the stretch of the Huron River within Ann Arbor city limits. Except carp, all are native to Michigan, although alterations in the river have changed both their local abundances and distributions. Agricultural and urban runoff, sewage discharges, and industrial pollution are some of the greatest causes of declining local water quality. Lower water quality makes the river unlivable for some fish species.

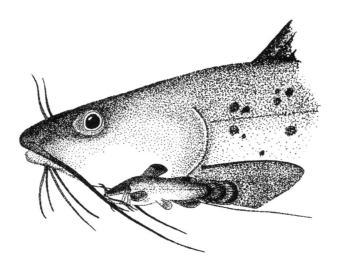

The channel catfish (larger) and the brindled madtom (smaller) represent the size extremes of fish inhabiting the Huron River in and around Ann Arbor.

One of the most dramatic alterations to fish habitat has been the construction of dams and the accompanying creation of impoundments. Dams block seasonal migrations and eliminate spawning areas for some fish species. They also increase water temperature, making conditions less than optimal for all but the hardiest fish. Still, impoundments provide recreational opportunities for park users, and do provide excellent habitat for some game fish, primarily bluegill and yellow perch. Barton, Argo, and Geddes Ponds are all impoundments along the course of the river which are popular fishing sites. Access for fishing is provided in most parks with river frontage. (State of Michigan fishing license is required.)

FISH SPECIES LIKELY TO BE FOUND IN THE HURON RIVER IN ANN ARBOR

Popular game species are in **bold** letters
Non-native species are indicated with an asterisk (*).

American brook lamprey
Banded killifish
Black crappie
Black redhorse
Black bullhead
Blackchin shiner
Blacknose dace
Blacknose shiner
Blackside darter
Blackstripe topminnow
Bluegill
Bluntnose minnow
Bowfin
Brindled madtom
Brook silversides
Brook stickleback
Brown bullhead
Carp*
Central stoneroller
Central mudminnow
Channel catfish
Common shiner
Creek chub
Fathead minnow
Golden redhorse
Golden shiner
Grass pickerel

Greater redhorse
Green sunfish
Greenside darter
Hornyhead chub
Johnny darter
Lake chubsucker
Largemouth bass
Longear sunfish
Longnose gar
Mimic shiner
Mottled sculpin
Northern pike
Northern hog sucker
Pumpkinseed sunfish
Rainbow darter
Rock bass
Smallmouth bass
Spotfin shiner
Spottail shiner
Stonecat
Striped shiner
Walleye
Warmouth
White sucker
Yellow bullhead
Yellow perch

The longnose gar is an example of a "living fossil." These fish have been on the earth for millions of years without going through significant evolutionary changes.

Amphibians

Frogs and Toads

Of the eight species of frogs and toads found in Ann Arbor during surveys conducted by volunteers in 1995, 1996, and 1997 six have been found in and along the Huron River. These six species are: spring peeper, chorus frog, American toad, gray tree frog, green frog, and bullfrog.

The **spring peeper**, Ann Arbor's most common frog, can be heard in full chorus starting in late March or early April. Its call heard singularly consists of a short high-pitched "peep", while a full chorus can sound like the jingling of sleigh bells. Spring peepers prefer temporary pools, marshes, and backwaters. Our volunteers have found them in the highest numbers near Barton and Ruthven Nature Areas. Peepers are small frogs which can be identified by their small toe-pads and a dark X on their back.

The spring peeper likes temporary pools, marshes, and backwaters

The **chorus frog** is similar in habit and habitat to the spring peeper, hence the two are often heard calling simultaneously in the warm evenings of April and May. The chorus frog call is a short trill which has been likened to running a thumbnail along the edge of a fine-toothed comb. This frog is quite small, usually with three dark lines running down the back and a white line above the lip.

Another resident of the Huron River corridor is the **American toad**, which has been found in most sites along the river including Bandemer, Furstenberg, Ruthven, and Barton Nature Areas and Gallup Park. It is easy to recognize due to its dry warty skin and brown color. The call of this toad is a long (up to 30 seconds) trill which is melodic but repetitive. The American toad prefers woodland habitats but is found in a variety of other habitats such as lake shores, meadows, and backyards.

Although the **gray treefrog** is one of our most widespread frogs in Ann Arbor, it prefers ephemeral shallow wetlands to the deeper water of the Huron River. Along the river, it has been found only in Ruthven Nature Area. This small frog can be identified by its textured skin and large toe pads (which are sticky, allowing it to easily climb trees). The call of the gray treefrog is a short, low-pitched nasal trill.

The green frog lives in the deeper water areas of the Huron River. This large frog sounds like a broken banjo string when it makes its calls in late summer.

The **green frog** is a frequent resident of the deep water areas common along the Huron River. It is a relatively late caller, getting its start in late May. Its call has been likened to the "twang" of a broken banjo string. The green frog is a large frog which can be distinguished from the bullfrog by two folds of skin running down its back. It has been found in Ruthven, South Pond, and Barton Nature Areas.

The **bullfrog** is uncommon within the city of Ann Arbor, and is only found along the Huron River. It is the largest frog in Michigan and ranges in color from yellow to green to brown. The call of the bullfrog is low and resonating and resembles a foghorn or the moo of a cow.

Currently, there have been no reports of the **leopard frog** in the river corridor of Ann Arbor, but this frog prefers open areas and may be a future visitor as efforts to restore such habitat continue with the NAP prescribed burn program. The call of the leopard frog is a low–pitched snore which has been likened to the sound resulting from rubbing a thumb along a balloon.

The **wood frog** prefers moist, shaded woodlands, and so far has been reported in wooded areas on the outskirts of Ann Arbor but not along the Huron River.

SEASONALITY OF FROG & TOAD CALLS IN THE ANN ARBOR AREA

	March	April	May	June	July	August
Wood frog	---------------------------					
Spring peeper	---					
Chorus frog	--					
Leopard frog		--				
American toad			------------------------------------			
Gray treefrog			---			
Green frog				---------------------------------		
Bullfrog				---------------------------------		

Salamanders

Visitors to the Huron River corridor are also likely to run into the closest relatives of the frogs, the salamanders. Of the 11 species found in Michigan, 6 are possible inhabitants of the Huron River corridor, although only 1 species, the mudpuppy, has been firmly documented in the recent past.

The mudpuppy is Ann Arbor's most surprising salamander when encountered. It can grow to over a foot in size and is permanently aquatic. This salamander retains gills associated with the larval phase for its entire life. Its color is drab with a lighter underbelly, and may be spotted in appearance. The mudpuppy can be differentiated from the larval stages of other salamanders by its hind feet— it has only four toes while other species have five. Mudpuppies inhabit rivers and lakes, spending much of their time foraging on the bottom. Mudpuppies have been declining in portions of Michigan due to their sensitivity to pollutants and chemicals used in fish management, and can be considered indicators of good water quality.

Other salamanders one might encounter are central and red-spotted newts, tiger salamanders, spotted salamanders, blue-spotted salamanders, and redback salamanders. Many of these species have been found in other areas of Ann Arbor, and perhaps are along the Huron River also. If present, they are most likely to be found in shallow, ephemeral ponds in spring, or under logs or leaf litter at other times of the year.

Reptiles

Turtles

Four turtles are commonly seen along the Huron River and nearby: the eastern painted turtle, Blanding's turtle, the eastern snapping turtle, and the spiny soft-shelled turtle. The painted turtle is very common in the ponds and marshes

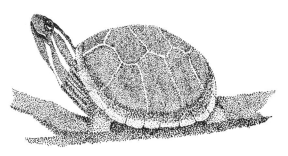

adjacent to the river as well as in the river itself. The sides of the throat are striped with yellow and red and the shell is dark with a criss-cross of yellow and finer red markings along the edge. Blanding's turtle is less common, but none-the-less often seen in the same habitats. It is easily identified by its yellow throat and yellow spotting on the shell. Also found in these same haunts is the larger, predatory, common snapping turtle. The shell of the

Painted turtles are very common in ponds and marshes adjacent to the Huron River. Note the characteristic stripes on the side of its throat.

snapping turtle has bony blunted spikes scattered across the top. Snapping turtles are essentially dull brown. Frequently only their triangular head, ending in a notched beak, is seen since these turtles prefer murkier water. Although essentially shy, the snapping turtle will bite a finger or toe placed too near its mouth. Finally, the spiny soft-shelled turtle is often seen along the Huron River, but rarely encountered elsewhere. This turtle is very flat in appearance, lacking the doming other turtles have. The shell, appearing very smooth, is olive overall with scattered bronzy-yellow spots. As with the snapping turtle, frequently all that is seen is the head, or the long tubular nose extending out of the water. Sometimes, however, when the water is clear the whole turtle is easily seen below the surface. The soft-shelled turtle will bite and should not be handled.

Snakes

Snakes are rarely seen along the course of the Huron River or in adjacent lands. With the decline of both the milk snake and fox snake, this area contains only two common species, the common garter snake and the northern water snake. These two species are easily distinguished from each other. The water snake has a series of dark brown bands separating lighter brown bands, and is indeed highly aquatic. The common garter

Garter snakes have long stripes running the length of their bodies. These may be green, blue, and sometimes red.

snake lacks the banding of the water snake. Instead it has long stripes running the length of the body. These can be in various shades of greens, blues, and sometimes reds.

Birds

Breeding Birds

The concept of an overall "bird nesting season" can be confusing because different species of birds nest during so many parts of the year— from late winter into early fall. The Great Horned Owl starts nesting as early as late January and the Mourning Dove nests as late as September or early October. Most species, however, nest from May through July and are found in many of the parks along the course of the Huron River (see tables on next two pages). These include many familiar species such as the cardinal, robin, House Wren, Song Sparrow, and many others.

During the summer breeding season more than 90 species of birds are regularly seen along the Huron River corridor, with most nesting locally. While many species are considered generalists (nesting in a wide variety of habitats), others have very specific nest–habitat requirements.

The Great Horned Owl starts nesting in the Ann Arbor area as early as late January. Look for this bird in forests with lots of snags (standing dead trees).

SUMMER RESIDENT BIRDS OF THE HURON RIVER CORRIDOR

	non-breeding resident	Barton	Bird Hills	Bandemer	Kuebler Langford	Argo	Cedar Bend	Nichols Arboretum	Gallup	Furstenberg	Ruthven	South Pond	Parker Mill
DOUBLE-CRESTED CORMORANT	x	x											
GREAT BLUE HERON	x	x		x		x		x	x	x		x	x
GREAT EGRET	x	x							x				
GREEN HERON	?	x				x		x	x	x		x	x
MUTE SWAN									x	x		x	x
CANADA GOOSE		x		x		x		x	x	x		x	x
WOOD DUCK		x	x	x		x		x	x	x	x	x	x
MALLARD		x				x	x	x	x	x		x	x
TURKEY VULTURE	[1]	x	x	x	x	x	x	x	x	x	x	x	x
COOPER'S HAWK		x	x					x	x	x			
BROAD-WINGED HAWK								x	x	x			
RING-NECKED PHEASANT		x						x		x	x		
VIRGINIA RAIL		x							x	x			
SORA		x							x	x			
KILLDEER		x		x				x	x	x		x	
SPOTTED SANDPIPER								x					
COMMON SNIPE		x											
AMERICAN WOODCOCK		x											
RING-BILLED GULL	x	x	x	x	x	x	x	x	x	x	x	x	x
ROCK DOVE		x	x	x	x	x	x	x	x	x	x	x	x
MOURNING DOVE		x	x	x	x	x	x	x	x	x	x	x	x
BLACK-BILLED CUCKOO		x	x					x		x			
YELLOW-BILLED CUCKOO		x	x					x					
EASTERN SCREECH OWL			x					x					
GREAT HORNED OWL								x		x			
COMMON NIGHTHAWK	x	x		x	x	x	x	x	x	x	x	x	x
CHIMNEY SWIFT	x	x	x	x	x	x	x	x	x	x	x	x	x
RUBY-THROATED HUMMINGBIRD		x	x				x	x	x	x			
BELTED KINGFISHER								x					x
RED-BELLIED WOODPECKER		x	x	x	x	x	x	x	x	x	x	x	x
DOWNY WOODPECKER		x	x	x	x	x	x	x	x	x	x	x	x
HAIRY WOODPECKER			x					x					
NORTHERN FLICKER		x	x	x	x	x	x	x	x	x	x	x	x
EASTERN WOOD-PEWEE			x					x		x			
ACADIAN FLYCATCHER			x					x					
WILLOW FLYCATCHER		x								x			
EASTERN PHOEBE		x							x				x
GREAT CRESTED FLYCATCHER			x					x		x			
EASTERN KINGBIRD		x						x	x	x			x
PURPLE MARTIN	x	x		x	x	x		x	x	x	x	x	x
TREE SWALLOW		x	x	x	x	x	x	x	x	x	x	x	x
NORTHERN ROUGH-WINGED SWALLOW		x						x	x	x		x	

[1] Nests only in Bird Hills

SUMMER RESIDENT BIRDS OF THE HURON RIVER CORRIDOR												
	non-breeding resident	Barton	Bird Hills	Bandemer	Kuebler Langford	Argo	Cedar Bend	Nichols Arboretum	Gallup	Furstenberg	Ruthven	South Pond	Parker Mill
BANK SWALLOW	x	x				x		x	x	x	x	x	x
CLIFF SWALLOW	[2]	x		x		x		x	x	x	x	x	x
BARN SWALLOW		x		x	x	x		x	x	x	x	x	x
BLUE JAY		x	x	x	x	x	x	x	x	x	x	x	x
AMERICAN CROW		x	x	x	x	x	x	x	x	x	x	x	x
BLACK-CAPPED CHICKADEE		x	x	x	x	x	x	x	x	x	x	x	x
TUFTED TITMOUSE		x			x		x	x		x			
WHITE-BREASTED NUTHATCH		x			x		x	x		x			x
CAROLINA WREN	x	x						x		x			
HOUSE WREN		x	x	x	x	x	x	x	x	x	x	x	x
MARSH WREN		x											
BLUE-GRAY GNATCATCHER		x	x	x			x	x		x			x
WOOD THRUSH		x						x					
AMERICAN ROBIN		x	x	x	x	x	x	x	x	x	x	x	x
GRAY CATBIRD		x	x	x	x	x	x	x	x	x	x	x	x
BROWN THRASHER		x								x			
CEDAR WAXWING		x	x	x	x	x	x	x	x	x	x	x	x
YELLOW-THROATED VIREO		x											
WARBLING VIREO	x			x	x	x		x	x	x	x	x	
RED-EYED VIREO			x					x					x
BLUE-WINGED WARBLER		x											x
YELLOW WARBLER		x				x		x	x	x	x	x	x
OVENBIRD			x		x			x					
COMMON YELLOWTHROAT		x		x	x	x		x	x	x		x	x
SCARLET TANAGER			x					x		x			
NORTHERN CARDINAL		x	x	x	x	x	x	x	x	x	x	x	x
ROSE-BREASTED GROSBEAK		x	x	x	x	x	x	x	x	x	x	x	x
INDIGO BUNTING		x						x		x			x
EASTERN TOWHEE		x	x				x	x		x			x
CHIPPING SPARROW		x	x	x	x	x	x	x	x	x	x	x	x
SONG SPARROW		x	x	x	x	x	x	x	x	x	x	x	x
RED-WINGED BLACKBIRD		x		x	x	x		x	x	x	x	x	x
EASTERN MEADOWLARK		x											x
COMMON GRACKLE		x	x	x	x	x	x	x	x	x	x	x	x
BROWN-HEADED COWBIRD		x	x	x	x	x	x	x	x	x	x	x	x
BALTIMORE ORIOLE		x	x	x	x	x	x	x	x	x	x	x	x
HOUSE FINCH		x	x	x	x	x	x	x	x	x	x	x	x
AMERICAN GOLDFINCH		x	x	x	x	x	x	x	x	x	x	x	x
HOUSE SPARROW		x	x	x	x	x	x	x	x	x	x	x	x

[2] Nests only in Bandemer and Gallup

Migratory Birds

The forests lining much of the Huron River and the creeks draining into the river serve as an important resting and feeding refuge for many species of migratory birds. These birds spend the summer breeding season in northern Michigan and Canada, moving south through our area to wintering grounds in the southern United States, Mexico, the Caribbean, and Central America. This group includes many warblers, vireos, flycatchers, and thrushes.

The spring is the time when many bird–watchers go out seeking these migrants, the season peaking in mid–May with numbers decreasing as June approaches. Many species are in full song and bright breeding plumage on their trip north, aiding greatly in their identification. The fall, however, is the time when the greatest number of migrants can be seen (those that migrated north the previous spring plus their offspring). Fall migration can sometimes begin as early as late July, but more typically migrants appear in greatest numbers in mid–August with many passing through until early October. Unlike the spring, few birds are singing in the fall, although many still have short, distinctive chip notes, and many have molted from bright breeding plumages to duller fall and winter ones. The lack of song and duller coloration increases the challenge of identification, but the greater numbers gives the careful observer many more opportunities.

For a summary of the common migrant birds of the Huron River corridor, see the table on the following page.

The Black–and–white Warbler migrates through Ann Arbor in April and September.

COMMON MIGRANT BIRDS OF THE HURON RIVER CORRIDOR

	Mar	Apr (Spring)	May (Spring)	June (Summer)	July (Summer)	Aug (Summer)	Sep (Fall)	Oct
SHARP-SHINNED HAWK		━	━			━	━	━
BLACK-BILLED CUCKOO*			━	━	━	━		
YELLOW-BILLED CUCKOO*			━	━	━	━	━	
OLIVE-SIDED FLYCATCHER			━			━		
YELLOW-BELLIED FLYCATCHER			━			━	━	
ACADIAN FLYCATCHER			━	━	━	━	━	
ALDER FLYCATCHER			━			?		
WILLOW FLYCATCHER*			━	━	━	━	━	
LEAST FLYCATCHER			━	━	━	━	━	
GREAT CRESTED FLYCATCHER*			━	━	━	━		
EASTERN KINGBIRD*			━	━	━	━		
TREE SWALLOW		━	━	━	━	━	━	
BARN SWALLOW			━	━	━	━	━	
RED-BREASTED NUTHATCH	━	━	━			━	━	━
BROWN CREEPER	━	━						
WINTER WREN		━					━	━
GOLDEN-CROWNED KINGLET	━	━	━				━	━
RUBY-CROWNED KINGLET		━	━					
BLUE-GRAY GNATCATCHER*			━	━	━	━		
WOOD THRUSH		━	━	━	━	━		
HERMIT THRUSH	━	━	━	━	━	━	━	━
CEDAR WAXWING*	━	━	━	━	━	━	━	━
SOLITARY VIREO		━	━				━	━
YELLOW-THROATED VIREO			━			━	━	
WARBLING VIREO*			━	━	━	━	━	
PHILADELPHIA VIREO			━			?		
RED-EYED VIREO*			━	━	━	━	━	
BLUE-WINGED WARBLER*			━	━	━	━		
GOLDEN-WINGED WARBLER			━			━		
TENNESSEE WARBLER			━			━	━	
NORTHERN PARULA			━			━		
YELLOW WARBLER*			━	━	━	━	━	
CHESTNUT-SIDED WARBLER			━					
MAGNOLIA WARBLER				━	━	━	━	
BLACK-THROATED BLUE WARBLER		━	━	━		━	━	
YELLOW-RUMPED WARBLER		━	━					
BLACK-THROATED GREEN WARBLER		━	━				━	━
BAY-BREASTED WARBLER			━				━	
BLACKPOLL WARBLER			━				━	
BLACK-AND-WHITE WARBLER		━	━			━	━	━
OVENBIRD*		━	━	━	━	━	━	
COMMON YELLOWTHROAT*			━	━	━	━	━	
CANADA WARBLER			━			━	━	
SCARLET TANAGER*			━	━	━	━	━	
INDIGO BUNTING*			━	━	━	━	━	
WHITE-THROATED SPARROW		━	━				━	━
BALTIMORE ORIOLE*			━	━	━	━	━	
PURPLE FINCH	━	━						━

Mammals

Along the course of the Huron River, adjacent woodlands, and old fields, you are likely to encounter a small number of mammals. This includes our only large mammal, the white–tailed deer. Others are best grouped by size for quick identification.

Medium
- Muskrat
- Mink
- Woodchuck
- Raccoon
- Opossum
- Eastern cottontail
- Striped skunk
- Red fox

Opossum are nocturnal animals found in virtually all natural communities in the Huron River corridor.

Muskrats appear as large wet rats with long hairless tails. They build numerous mounds along the river and in the marshy wetlands. These mounds serve not only as their homes, but also as nesting platforms for Canada Geese. Mink are as aquatic as the muskrat, and are quite predatory. They are easily distinguished from the muskrat by their long-bodied profile, thick, bushy tail, and whitish–tan chin. The woodchuck, also called groundhog, is the least aquatic of the mid–sized mammals, preferring large fields and gardens. It is a stocky animal with short legs and a short, bushy tail. Raccoons, opossum, cottontails, and striped skunks are found in almost all habitats in the area. The red fox, on the other hand, exists in small numbers at scattered locations such as Bird Hills, Furstenberg, and Barton Nature Areas.

Small
- Fox squirrel
- Eastern gray squirrel
- Red squirrel
- Thirteen–lined ground squirrel
- Southern flying squirrel
- Eastern chipmunk

The fox squirrel is the most common of the group, and inhabits most urban forested areas in the Huron River corridor. The gray squirrel is far less common than the fox squirrel and is most easily found in the larger forests along the river, and at a few other locations around Ann Arbor. The red squirrel is dependent on conifers. There is a large population of red squirrels along the river in and near Nichols Arboretum. The thirteen–lined ground squirrel is one of the least common of this group, found only in large open fields in the area. The status of the southern flying squirrel is unclear since few are seen because of their nocturnal habits. Chipmunks, of course, can be seen thoughout the corridor.

Very Small
- Moles (Eastern, Star–nosed)
- Shrews (Masked, Least, Northern short–tailed)
- Voles (Meadow, Woodland)
- Mice (Deer, White–footed)
- Bats (Little brown, Northern, Indiana, Red, Hoary, Silver–haired, Big brown)

White–footed mouse

Meadow vole

Very few of these are ever seen, fewer identified. This is because of their nocturnal habits and rapid retreat from humans. Both the big brown bat and the little brown bat are fairly common in the area, although they are rarely observed during daylight and are very difficult to distinguish in flight.

Unusual or rare
- Ermine, or Short–tailed weasel
- Long–tailed weasel
- Least weasel
- American badger
- Coyote

These animals are rarely, if ever, encountered along the Huron River corridor in Ann Arbor.

THE NATURAL AREAS

The following pages describe each of the 13 natural areas along the Huron River corridor in much greater detail. Maps of individual areas are also included in this section. The maps delineate the natural communities in each park, as well as other information such as park boundaries, trails, and parking areas.

The natural areas are arranged alphabetically, so refer to the large fold–out map inside the back cover of the guide to locate each park relative to the others. All information provided is current at the time of this writing. Site names use both "Park" and "Nature Area" for properties owned by the City of Ann Arbor.

The natural areas of Ann Arbor include this scene of the Huron River oxbow at Barton Park. The following pages provide a guide to the different natural areas of the Huron River corridor, including maps, information on natural communi–ties, and information about services at the different parks.

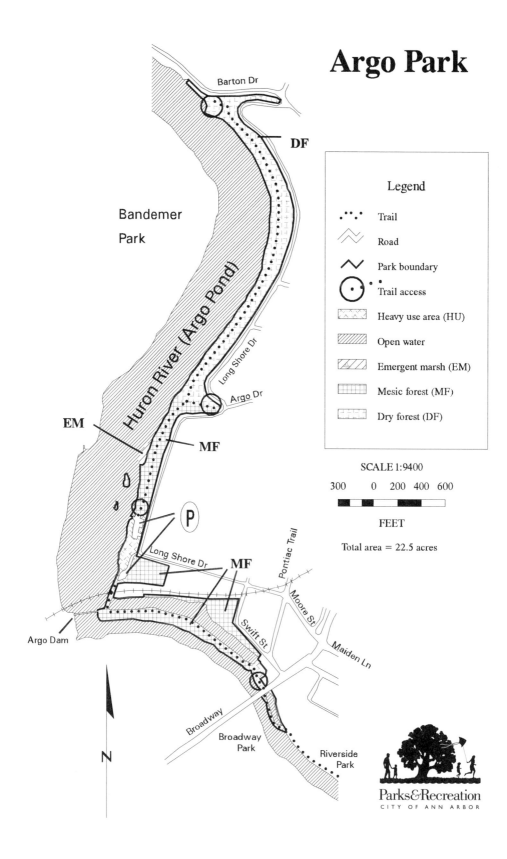

Argo Park

Barton Dr

DF

Bandemer
Park

Huron River (Argo Pond)

Long Shore Dr

Argo Dr

EM

MF

P

Long Shore Dr

MF

Pontiac Trail

Moore St

Swift St

Maiden Ln

Argo Dam

Broadway

Broadway
Park

Riverside
Park

N

Legend

⋯	Trail
〰	Road
︿	Park boundary
⊙	Trail access
▨	Heavy use area (HU)
▨	Open water
▨	Emergent marsh (EM)
▦	Mesic forest (MF)
▥	Dry forest (DF)

SCALE 1:9400

300   0   200 400 600

FEET

Total area = 22.5 acres

Parks&Recreation
CITY OF ANN ARBOR

ARGO NATURE AREA

Parking	east of Longshore Drive, adjacent to the Argo canoe livery
Trail Access	parking area, shore of pond
Restrooms	none
Picnic Areas	picnic table
River Access	numerous fishing sites along shore of Argo pond
Comments	22.5 acres; canoe livery and launch

General Information

Argo Nature Area is located along Longshore Drive on the east shore of the Huron River. This stretch of the river contains an impoundment created by Argo Dam which is also referred to as Argo Pond. Most of Argo Park is north of the dam and the canoe livery, but there is also a section below the dam, running southeast along the spillway and extending to Broadway Street. An unpaved trail runs the entire length of the park, from Barton Drive to the north, to Broadway Street and Riverside Park to the southeast. The northern section of trail offers spectacular views of Argo Pond at several openings in the forest.

North of the dam is a **canoe livery**, where visitors can rent canoes in which to explore Argo Pond and the rest of the river. (Call 668-7411 between April and October for more information.) Canoeists may paddle around the east end of the dam via the spillway which ends in a short carryover to the main channel of the river. There is a small snack bar in the livery.

History

Argo Park is one of Ann Arbor's oldest river parks. In the 1930s and 1940s, Argo housed the only public bathing beach in Ann Arbor. It was located where the canoe livery is today. When declining water quality became a problem, the beach was closed. Twenty years later, the city bought the land from Detroit Edison, which was using Argo Dam as a power source.

Future Plans

Future plans for the Argo Park area include a boardwalk along Barton Drive to connect it to the Bandemer Park entrance west of M–14. This trail will then continue to Barton Park to allow a long and scenic continuous walk along the river. Construction on this trail network is scheduled to begin in 1998.

Natural Communities

The three natural communities found in Argo Nature Area are mesic and dry forest, and emergent marsh. In the 1994–1998 plant inventory, 293 species of plants were recorded here, 200 of them native.

About one–half of Argo Nature Area is **mesic forest**. Some of the more common trees you will encounter along the trails here include American basswood (*Tilia americana*), crack willow (*Salix fragilis*), and hornbeam (*Carpinus caroliniana*). Two of the more unusual flowers in Argo Nature Area are green dragon (*Arisaema dracontium*), and monkey-flower (*Mimulus ringens*). Both plants grow in moist, shaded areas.

Like most of our mesic forests, the one in Argo Nature Area is being overrun with invasive shrubs, especially honeysuckle (*Lonicera*) and buckthorn (*Rhamnus*). One of the factors contributing to this invasion in Argo is the shape of the park. Argo is a long narrow strip of land, with no point further than 100 feet from the edge of the forest. This enables birds feeding on berries outside of the park to fly into the interior of the forest and disperse the seeds, something they couldn't easily do if it were a large, circular block of forest canopy.

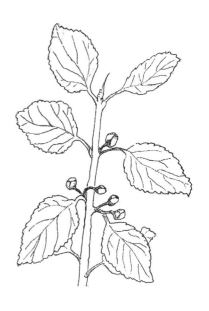

A second factor which may contribute to the spread of invasive shrubs in Argo is the severe erosion which occurs along the trail north of the canoe livery. The slope here is naturally very steep because it is on the outside edge of a bend in the river and thus subject to erosion (See the *Reading the Landscape* section for a full description of this process). Since the construction of Argo Dam, however, some of this erosive action has been reduced.

Buckthorn (Rhamnus) shrubs are some of the most troublesome invasives in Ann Arbor's natural areas. Their abundant black berries attract birds which then spread the seeds into other natural areas.

A certain amount of erosion on steep river banks is natural, but there have been other factors at work in Argo. Runoff from Longshore Drive periodically cascades off the edge of the road, channeling rainwater across the forest floor and carving erosion gullies. This disturbance to the forest soil provides an ideal germination site for invasive plants.

Compaction of the soil by hikers or bikers also increases the amount of runoff because rainwater is less able to percolate into the soil. Because of these concerns, and because of the narrowness of the trail, bicycles are not allowed north of the canoe livery and hikers are asked to stay on the trails.

In addition to the widespread invasive shrubs, the mesic forest on the steep slopes of Argo Nature Area is home to some less common non–native plants. Wild stonecrop (*Sedum ternatum*) is native to parts of eastern North America but probably not to Washtenaw County. It is known in Ann Arbor only from Argo Nature Area; the plants here may have been introduced either deliberately or from someone's garden waste and have spread into a sizable colony. It is a ground cover plant that grows on the dry slopes along the river. Along the trail southeast of the dam you may notice a clinging vine with multiple leaflets radiating from a single point. This is chocolate–vine (*Akebia quintata*), a non–invasive exotic which is also unique to Argo. Let's hope that neither of these exotics becomes too widespread in the future.

The short spur trail to Argo Drive marks the border of the mesic forest to the south and the **dry forest** to the north. Standing at this line, one can appreciate the difference between the dense shrub layer in the moist forest and the relatively open understory of the drier site. As you walk north, look for the block–like bark of the large black oaks (*Quercus velutina*) growing here. The thickness of the bark protects the tree from the potentially harmful effects of fire. Also scattered throughout the woods are pignut and shagbark hickories (*Carya glabra* and *Carya ovata*). While the bark on both of these tress can be

Black oak leaves

Large black oaks (*Quercus velutina*) are scattered through the dry forest section of Argo Park. Like red oaks, the leaves of black oaks have bristled tips. However, the bark on black oaks is more block–like than that of the red.

quite flaky, the shagbark's arches well away from the tree, while the pignut's doesn't curl quite as prominently. Look for patches of American hazelnuts (*Corylus americana*), whose fruit is prized by squirrels, and the early winter yellow flowers on witchhazel's (*Hamemelis virginiana*) arching branches.

Aside from the mesic and dry forest, the only other natural community in Argo Nature Area is a small patch of **emergent marsh** along the bank of Argo Pond. This site is easily accessible only by canoe.

Other Wildlife
Argo Pond is one of the few sites in the city where bullfrogs were recorded in the 1996 survey. Also in 1996, two uncommon butterflies were found in the park: the Harvester (summer and fall) and Leonard's Skipper (fall). Screech owls were reported nesting here in 1994.

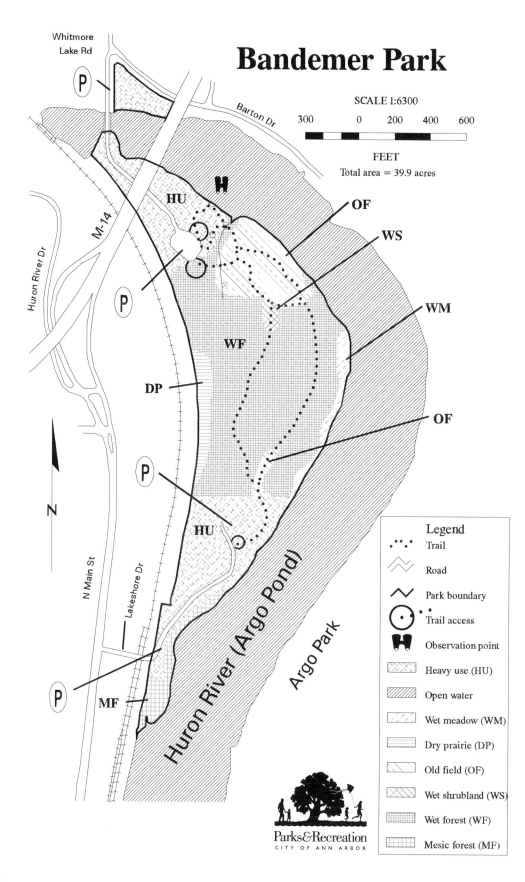

BANDEMER NATURE AREA

Parking	1) main vehicle lot at end of north entrance drive, 2) at north entrance on Whitmore Lake Road, 3) at Lakeshore Drive entrance; bicycle rack at main lot
Trail Access	parking lots
Restrooms	portable ones near north end
Picnic Areas	tables and grills at north end of park; picnic shelter
River Access	handicap accessible canoe dock at north end; (rowing dock at south end is private)
Comments	36.9 acres; artesian well

General Information

Bandemer Nature Area is located on the west shore of the Huron River next to Argo Pond. The entire length of the park is bounded on the west by Conrail railroad tracks near North Main Street. Construction at the more–manicured north end of the park was completed in 1996 and features paved and gravel trails. Interior trails are unpaved. The southern end of Bandemer is the current (1998) location for several businesses whose leases will expire in the next several years. Several area rowing clubs also lease a building here for boat storage. The long dock at this end is for use by these rowing clubs.

History

The name for the park comes from Ted and Mary Bandemer, who donated funds to purchase the northern portion of the property in 1985. Historically, this part of the park was used as pasture land and as stockyards for the railroad. At one time, much of the interior of the park was used as a dumping ground; first when it was owned by a construction company and then when it was sold for the construction staging of M–14. Approximately one-fifth of Bandemer is actually old river channel that was filled in by previous owners. As a result of these disturbances, many of the plants here are exotic invasives.

Long before humans dumped soil here, the river was doing the same thing. As the river cut into the steep bank of Argo Nature Area on the opposite shore, it also dropped sediments in Bandemer, adding to the glacial outwash plains started 14,000 years ago. (See the *Reading the Landscape* section for a more complete explanation of these processes.)

Future Plans

Bandemer Nature Area is still evolving. As the leases for resident businesses expire, the city will consider whether each building has a future use, or whether it should be removed. Future plans call for linking Bandemer to Barton Nature

Area via a direct trail connection under the railroad tracks at the trestle over the river.

Natural Communities

Despite the historical abuses to the land, Bandemer still contains a wide variety of natural communities due to the extreme moisture gradient across the site. These communities include: dry prairie, old field, wet forest, wet shrubland, mesic forest, and wet meadow. In the 1994–1998 plant inventory, 275 species of plants were recorded here, 184 of them native.

Along the railroad on the west edge of the park is an area best described as **dry prairie**. Although much of the ground here is covered with a thin layer of gravel and slag from the railroad, you can still find big bluestem grass (*Andropogon gerardii*), switchgrass (*Panicum virgatum*), and other prairie species thriving in the impoverished soil. This area was burned in spring, 1996 to discourage the encroaching woody growth.

The **old field** in the interior of Bandemer has similarities to the dry prairie, but it lacks the brilliant fall colors you'll see along the railroad tracks. Like the prairie, however, this is a good place to look for butterflies in the summer. The old field here is dominated by non-native species such as Kentucky bluegrass (*Poa pratensis*), Canada bluegrass (*Poa compressa*), and spotted knapweed (*Centaurea maculosa*). As you walk through this site, beware of the sharp spines of the native hawthorn shrubs (*Crataegus*) growing here.

The majority of the park is classified as **wet forest**, which is becoming established on most of the fill soil. A high water table and large amount of clay in the fill have resulted in pockets of standing water throughout the forest. The invasive common buckthorn (*Rhamnus cathartica*) quickly colonized this site and dominates it now. Visitors can get a sense of how aggressive this shrub can be by walking either of the north–south trails through the heart of the park and observing its dominance. If you walk the western trail, look for the huge old American elm tree (*Ulmus americana*) 50 yards to the east; its broad spreading branches are a monument to a time when this area was open and sunny. Enjoy this tree from a distance, however; it is dead and occasionally sheds its large limbs.

Just north of this giant elm, on an especially wet pocket of fill, the trail leads you through a tiny patch of **wet shrubland**. This area is muddy year-round, although trail improvements are planned to help keep your feet dry. Glossy buckthorn (*Rhamnus frangula*) forms a dense thicket in this community.

Although it is actually closer to the river, the **mesic forest**

*The large orange flowers of Michigan lily (*Lilium michiganense*) may be spotted in the wet meadow community of Bandemer Park.*

is slightly drier than the wet forest because it does not grow on the clay–rich fill. Look for this thin strip of forest along the river at the southern entrance to the park.

Further north along the river, a **wet meadow** survives at the water's edge. From a canoe is the easiest way to see the odd–shaped flowers of turtlehead (*Chelone glabra*), the large orange flowers of Michigan lily (*Lilium michiganense*), or the dark purple flowers of marsh pea (*Lathyrus palustris*), a member of the bean family. Also look for abundant Missouri ironweed (*Vernonia missurica*) in this area.

Other Natural Features
Another natural community is being established in a very unnatural portion of Bandemer Park. Directly below M–14, a **wetland** is being created which will catch the runoff water from the highway and help clean it before it enters the river. The harshness of this site makes this a challenging project: it is in constant shade, salty water comes pounding down from the bridge, and highway crews require periodic vehicular access to the site. Watch this project develop over the next few years.

Finally, Bandemer provides an excellent opportunity to view an interesting feature found at several locations along the river: an **artesian well**. It is located north of the rowing clubs' dock between the gravel road and the river. An artesian well is simply groundwater under pressure. It is formed when an impermeable soil layer, such as clay, keeps the groundwater from rising to the surface. Pressure is created when the groundwater that feeds the well is located at a higher elevation. If the impermeable seal is punctured—in this case, with a pipe—the water comes bubbling up from the ground. For this artesian well, pressure probably comes from groundwater in the nearby bluffs on the west side of North Main Street. Caution: this water has not been tested and should not be consumed!

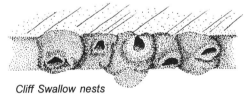

Cliff Swallow nests

Other Wildlife
Bandemer is home to the Harvester, an uncommon butterfly in Michigan. The Harvester is found throughout this stretch of the Huron River where the shoreline is dominated by the invasive black alder tree (*Alnus glutinosa*). (See the section on *Butterflies* to find out the connection.)

Bandemer is also home to a small colony of Cliff Swallows, which are uncommon birds in southern Michigan. These birds build their nests under the M–14 bridge high above the Huron River.

Finally, in spring, Bandemer Nature Area is an excellent place to go to hear the long trill of American toads, which are quite abundant here.

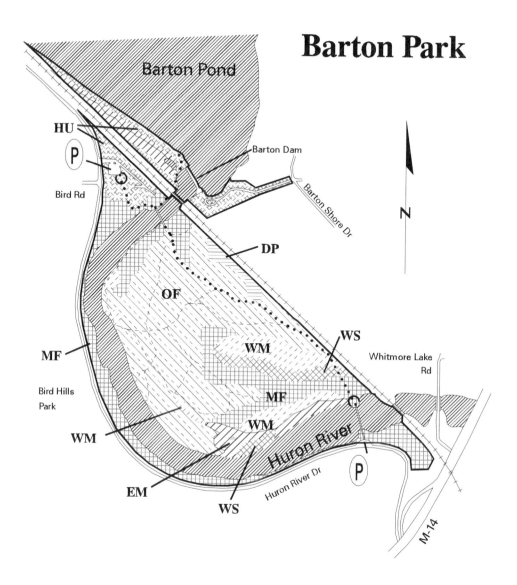

Barton Park

Barton Pond

HU

P

Bird Rd

Barton Dam

Barton Shore Dr

DP

OF

N

WS

Whitmore Lake Rd

MF

Bird Hills Park

WM

MF

WM

Huron River

WS

EM

P

Huron River Dr

M-14

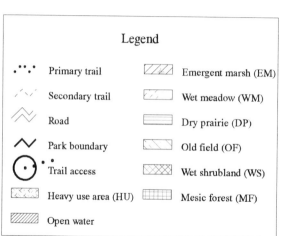

Legend

•⋯•	Primary trail	▨	Emergent marsh (EM)
⌇	Secondary trail	▨	Wet meadow (WM)
⋀	Road	▤	Dry prairie (DP)
⋀	Park boundary	▨	Old field (OF)
⊙	Trail access	▨	Wet shrubland (WS)
▨	Heavy use area (HU)	▥	Mesic forest (MF)
▨	Open water		

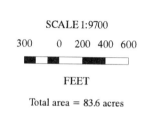

SCALE 1:9700

300   0   200 400 600

FEET

Total area = 83.6 acres

Parks&Recreation
CITY OF ANN ARBOR

BARTON NATURE AREA

Parking	1) main lot is on Huron River Drive near Barton Dam, 2) pull-off a mile further east, near M-14
Trail Access	either parking area
Restrooms	a portable one near the main parking lot
Picnic Areas	tables and grills near the main parking lot, table near the dam
River Access	canoeists have a short portage around the west end of the dam; numerous fishing sites along riverbank, especially near dam, where there is also a small dock in Barton Pond
Comments	83.6 acres

General Information

The main part of Barton Nature Area is the flat, open shrubby area visible across the river from Huron River Drive just below Barton Dam. There is one official trail connecting the two bridges; it is eight feet wide and covered with woodchips. There are also numerous unofficial paths leading down to the river from various points along this trail. Near the dam, another path leads from the bridge to Barton Pond, and then on top of the earthen dam to the northwest toward Foster. Although Foster is officially a part of Barton Nature Area, this guide refers to Foster as a separate park because of its distinct location and ecological qualities.

Locally, some people know Barton as "Oxbow Park." What is an "oxbow"? An oxbow is a circular bend in the river created as the river meanders over time. What starts as a gentle curve, eventually grows into a sharper and sharper arch as the river continues cutting further into the outside bank. If the river snakes around far enough, it may even loop back into itself, reconnect, and cut off part of the former bend. The isolated C-shaped lake is called an "oxbow lake." Technically, therefore, the term "oxbow" refers to the bend in the river and not the adjoining land. (See the section on *Reading the Landscape* for a complete explanation and diagram of this process.)

History

Barton Nature Area was purchased in the 1960s from Detroit Edison, who used Barton Dam for electric power production. Barton is the only dam in the city still used to generate electricity, but now the City sells the electricity to Detroit Edison. Historically, the flat nature of this land made it attractive for farming. The bridges and the wide woodchip trail were constructed in 1991 to supplement the informal network of trails which has evolved over time.

Natural Communities

The oxbow area in Barton Nature Area is a patchwork of natural communities, including: old field, dry prairie, wet meadow, wet shrubland, mesic forest, and emergent marsh. In the 1994–1998 plant inventory, 314 species of plants were recorded here, 235 of them native.

The majority of the oxbow area is classified as **old field**. Clues from remnant plants suggest that it used to be prairie, but became an old field after being abandoned as farmland. Still, there are many native wildflowers to enjoy. The large blue–pink flowers of robin's plantain (*Erigeron pulchellus*) are especially striking in the early spring. The fragrant flowers of bee-balm (*Monarda fistulosa*), the bright yellow flowers of black-eyed Susan (*Rudbeckia hirta*), and the white-flushed–with–violet flowers of hairy beard–tongue (*Penstemon hirsutus*) highlight the summer. In the fall, the brilliant yellow color of showy goldenrod (*Solidago speciosa*) lights up the landscape.

*Black–eyed Susan (*Rudbeckia hirta*) is commonly found in old field communities. Look for the bright yellow flowers on stalks 2–3 feet tall growing in open sunny areas of Barton.*

In spring or fall NAP staff burns areas of this park to drive back the hawthorn (*Crataegus*), buckthorn (*Rhamnus*), and honeysuckle (*Lonicera*) shrubs which were closing in on these sun–loving wildflowers. Well into the future, the effect of this and future burns will be apparent in the field of dead, or at least charred, shrubs visible from the main trail.

Closer to the railroad, the old field has recovered to what can once again be called **dry prairie**. The main distinction here is the thick stand of big bluestem grass (*Andropogon gerardii*) which dominates in patches. This is the densest stand of this six–foot high grass that you will find in any of our city parks. Big bluestem is probably the most widespread of any of the tall prairie grasses. It was once quite abundant in the tall–grass prairies which stretched for hundreds of miles across the Midwest. Catch it at its peak of color and height in September and October.

The proximity of this prairie remnant to the railroad tracks is not mere chance. Sparks thrown from passing trains would have ensured that this area historically burned with enough regularity to keep out the encroaching shrubs.

As the river worked its way out to its current path with ever-widening swaths, it left behind a few former channels that are slightly lower and wetter than the old field and dry prairie. Here the **wet meadow** community dominates. These rich areas are home to many birds and butterflies, as well as numerous moisture–

loving plants such as sneezeweed (*Helenium autumnale*), the pale pink, flat–topped flowers of Joe–pye weed (*Eupatorium maculatum*), and numerous sedges (*Carex*). Also look for the less common tufted loosestrife (*Lysimachia thyrsiflora*) and yellowish flowers of swamp–betony (*Pedicularis lanceolata*).

Like prairies, wet meadows are also maintained by fires. Where no fires have occurred recently, the wet meadows have been invaded by shrubs to the point where they are now classified as **wet shrubland**. Red–osier dogwood (*Cornus stolonifera*), elderberry (*Sambucus canadensis*), and willow (*Salix*) are the most common native shrubs here, through there are also many non–natives here. Identify red-osier dogwood by its bright red stems. This is also the only known site in Ann Arbor for another interesting plant, wild senna (*Cassia hebecarpa*). This tall, striking plant grows to 5 feet in height, and is topped with a large cluster of bright yellow butterfly–like flowers.

One of the densest stands of big bluestem (Agropogon gerardii) is in the northeastern portion of Barton Park near the railroad tracks. It is at its peak in color and height in September and October.

Woody plants draw moisture out of the soil and release it to the atmosphere as they "breathe." This can have the effect of drying the soil and lowering the water table. This may be what has allowed a patch of **mesic forest** to replace the wet meadows and wet shrublands in part of the oxbow area. Tall agrimony (*Agrimonia gryposepala*), enchanter's nightshade (*Circaea lutetiana*) and Jack–in–the–pulpit (*Arisaema triphyllum*) are some of the wildflowers which can be found beneath basswood (*Tilia americana*) and American elm (*Ulmus americana*) trees.

A separate strip of mesic forest runs along Huron River Drive on the *other* side of the river. Look for the tall, straight cottonwood (*Populus deltoides*) and the much less common butternut (*Juglans cinerea*) trees in this stretch. Two of the common low–growing spring wildflowers here are spring cress (*Cardamine bulbosa*) and wild ginger (*Asarum canadense*).

At the edge of the river, along the backwater of several small bays, an **emergent marsh** can be found. In the spring, canoeists may spot blue flag iris (*Iris virginica*), water smartweed (*Polygonum amphibium*) or marsh marigold (*Caltha palustris*). Enjoy them while you can; in the summer the scene becomes dominated by purple loosestrife (*Lythrum salicaria*).

Plant life is also present underneath the surface of the water at the river's edge. Canoeists can feel these submerged plants with their paddles and look down to see coontail (*Ceratophyllum demersum*), pondweed (*Potamogeton*), great bladderwort (*Utricularia vulgaris*), and water-milfoil (*Myriophyllum*) growing there.

Animals

The oxbow area of Barton Nature Area is home to several butterflies that are uncommon in the area. American and Bronze Coppers are found in the open fields, while the state–threatened Wild Indigo Duskywing can be seen among patches of crown vetch (*Coronilla varia*). Recently, the Little Sulphur and Orange Sulphur, both southern species, have been located in the park.

The brushy thickets adjacent to the prairie and old field are home to the Brown Thrasher, an otherwise rare bird in the Ann Arbor area. According to *The Birds of Washtenaw County, Michigan*, this is one of the few locations in Washtenaw County where over–wintering robins can be seen. Canada geese, various ducks, and Great Blue Herons are often visible from the southern bridge.

Listen for the calls of chorus frogs and spring peepers in Barton Park. Red fox are sometimes seen in the park, and muskrats are common along the river banks.

Most Great Blue Herons arrive in the Huron River corridor area in early spring and stay through the end of August.

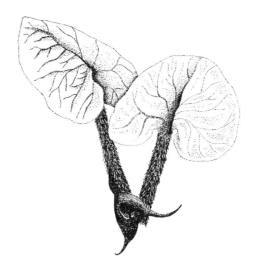

Wild ginger (Asarum canadense) grows in mesic forests in Barton Park. Look for its bright green heart-shaped leaves growing close to the ground.

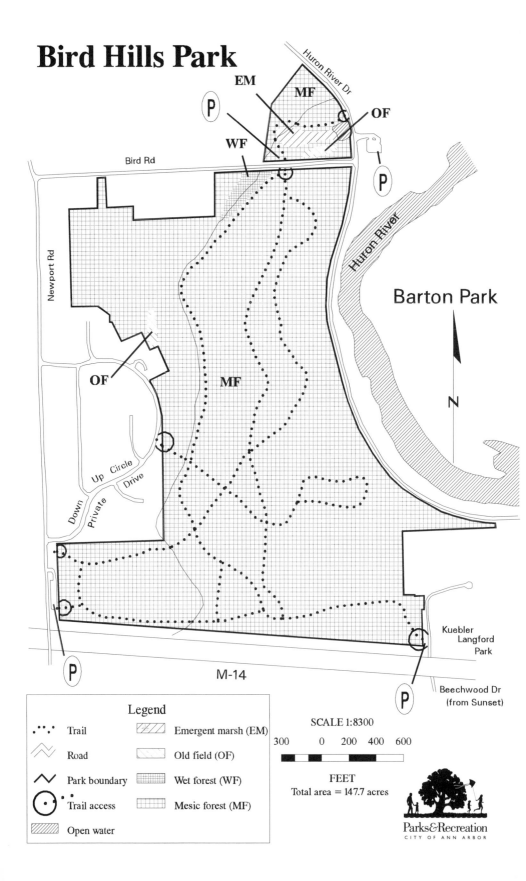

BIRD HILLS NATURE AREA

Parking	1) main lot on Newport Rd, 2) pull-off on Bird Rd and 3) Beechwood Dr (off sunset); bicycle racks at each lot
Trail Access	all parking lots and from Down-Up Circle
Restrooms	none
Picnic Areas	none
River Access	none
Comments	147.7 acres; farm remnants; County champion sugar maple; interpretive kiosk

General Information

Bird Hills Nature Area is located on the west side of Huron River Drive, across the river from Barton Nature Area. It is the largest park in the city, with several miles of unimproved trails running up and down its hilly interior. Bicycles are not allowed on any of the trails in Bird Hills. An interpretive brochure is available at an information kiosk at the main parking lot, or from the Parks and Recreation office in the Guy C. Larcom, Jr. Municipal Building.

The hilly nature of Bird Hills Park is due to its location on the Fort Wayne moraine. (See the section on *Reading the Landscaping* for further explanation.) There is a 176–foot elevation change within the park.

History

The property was logged in the late 1800s. After the trees were removed, the area was farmed; some remnants of the farm are still visible in the park. Running south from Bird Road are two lines of parallel concrete tracks. They mark the site of a farm road designed to fit the wheels of a wagon while letting the horse or ox pulling the wagon walk on softer ground between the tracks. Look for large, spreading, open-grown trees which probably grew up in fence rows during the time the area was maintained as a pasture.

In the early 1900s, the Graves family bought the property as a potential development site. At this time much of the main ridge in the park was still cattle pasture and was nearly treeless. The absence of plant cover increased the amount of runoff from the high ridge after a rain. This caused severe erosion in the steep ravines running down to the river. The interpretive kiosk and brochure provide photos showing the striking differences in the Bird Hills landscape between then and now.

To make the land more appealing for development, Henry Graves planted a variety of trees on the open site. This contributes to the diversity of woody plants in Bird Hills today, and explains why many non–native trees such as Douglas fir (*Pseudotsuga menziesii*), Scots pine (*Pinus sylvestris*), and white fir (*Abies concolor*) can still be found there.

In 1967, the City bought the majority of the land from the Graves family to be used as a park. In the 1970s, when plans to create a condominium complex next to the park surfaced, neighbors and other citizens rallied and raised money to help buy the land. Again in 1990, more land adjacent to the now–larger park was threatened by development, and again the citizens and City teamed up to buy the land to add to the park.

The tiny triangle of the park north of Bird Road has its own interesting history. In the 1920s, it was the site of one of the main wells for the city's drinking water. Later it was used as a police pistol range.

Natural Communities

While Bird Hills contains four distinct natural communities, mesic forest, old field, wet forest, and emergent marsh, the vast majority of the park is classified as mesic forest. The communities at Bird Hills are rich, with 100 woody species and 358 total plant species found here, 278 of them native, including 5 which are listed as either endangered or special concern in Michigan!

Mesic forest now covers most of the park, including the ridge-top which was pasture earlier this century. But the richest sections of the forest are on the eastern slope of the ridge, in a series of ravines leading down to Huron River Drive. These ravines were probably widened and deepened by the excess runoff from the pastures. Many of these ravines are watered by seeps, or springs, coming out of the slope. These seeps are formed when rainwater percolating down through more porous soil hits an impermeable layer and follows that layer horizontally to the side of the slope. This creates a moist

The mesic forest in the southern half of Bird Hills supports one of the city's most dramatic displays of flowering dogwood (Cornus florida) in the understory of the forest canopy.

percolating down through more porous soil hits an impermeable layer and follows that layer horizontally to the side of the slope. This creates a moist environment which supports a diversity of ferns such as maiden hair fern (*Adiantum pedatum*), whose graceful fronds arc over the forest floor, and fragile fern (*Cystopteris protrusa*), a small fern found in no other park in Ann Arbor. Other wildflowers found in the mesic forest of Bird Hills include blue cohosh (*Caulophyllum thalictroides*), early meadow–rue (*Thalictrum dioicum*), and wood sage (*Teucrium canadense*).

These ravines are also wonderful habitats for trees such as beech (*Fagus grandifolia*) and sugar maple (*Acer saccharum*). In fact, the largest sugar maple in the county grows here, with a 59.1–inch diameter! These native maples are being threatened by the invasive Norway maple (*Acer platanoides*) which is reproducing profusely in parts of the park. Although the leaves of young Norway maples are similar to those of sugar maple, they can be easily distinguished by the milky–white sap which oozes from a broken petiole, the long stem that connects the leaf to the twig.

The southern half of Bird Hills is also mesic forest and supports one of the largest spring displays of flowering dogwood (*Cornus florida*) in the city. Equally brilliant are the carpets of large–flowered trillium (*Trillium grandiflorum*), wild geranium (*Geranium maculatum*), and May apple (*Podophyllum peltatum*) found every spring.

Just as expansive in Bird Hills is the carpet of the invasive ground cover called myrtle or periwinkle (*Vinca minor*) which is taking over parts of the park, especially its western and eastern slopes. The periwinkle-blue flowers and shiny green foliage make it an attractive plant which has been widely planted in shady yards or flower beds. Unfortunately, as seen here, it can escape and invade our natural areas.

At the north end of the park, a **wet forest** community runs along either side of the small intermittent stream that flows north and eventually crosses Bird Road. Here you can find skunk–cabbage (*Symplocarpus foetidus*), golden ragwort (*Senecio aureus*), and the delicate orange flowers of spotted touch–me–not (*Impatiens capensis*). When the tiny bean–like seed pods of touch–me–not are ripe in late summer, they will explode at the slightest touch, scattering seeds in all directions.

Some uncommon plants that are found in the forests of Bird Hills include: bishop's cap (*Mitella diphylla*), a delicate plant which has a single flowering stalk with only two leaves; sharp–lobed hepatica (*Hepatica acutiloba*) which has broad pointed leaves and bluish flowers; richweed (*Collinsonia canadensis*) whose cluster of summer–blooming pale yellow flowers is borne over broad toothed leaves; and false spikenard (*Aralia racemosa*), a large, robust plant which can be found in the protected ravines of Bird Hills. This is also one of the parks where the rare butternut tree (*Juglans cinerea*) grows. Of special interest here is a shrub found only in Bird Hills and the adjacent Kuebler–Langford Nature Area: spicebush (*Lindera benzoin*), whose leaves smell spicy when crushed.

Follow the intermittent stream north to the other side of Bird Road and you'll discover small examples of additional natural communities. An **old field** runs close to the border of the park. While this area is quickly becoming overgrown with trees and shrubs, you can still find the cotton–like heads of thimbleweed (*Anemone virginiana*) growing beside the pale purple fall–flowering smooth aster (*Aster laevis*). As you walk through the old field remnant, note the spread of black locust (*Robinia pseudoacacia*). While this plant is native to the southeastern United States, here in Michigan it spreads quite aggressively, largely by shoots that sprout up from the roots. If you leave the trail and walk

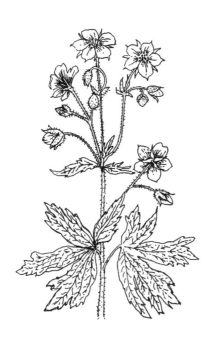

up the small hill to your right, you enter a higher–quality old field remnant, one that could also be classified as dry forest. The rapid change in plants around you marks the abrupt transition into this small natural community. Little and big bluestem (*Schizachyrum scoparium* and *Andropogon gerardii*) wave gently in the breeze below the drooping branches of the dry soil-loving northern pin oaks (*Quercus ellipsoidalis*). Another interesting native plant is the pasture thistle (*Circium discolor*). While this thistle is just as prickly as its numerous non–native cousins, the white undersides of the leaves provide an excellent distinguishing characteristic of native thistles.

Wild geranium (Geranium maculatum) is rose–purple in color and commonly grows in the understory of mesic forests. Look for the distinctive deeply–lobed, lacy leaves.

The remaining natural community is just downhill from the old field remnant. An **emergent marsh** forms a geographical transition between the upland portions of the park and the mesic forest further north. Growing close to the ground, the clear blue flowers with golden centers are forget–me–nots (*Myosotis scorpioides*), frequent escapees from local gardens. Shading these delicate flowers are the simple leaves of the wet–loving silky dogwood (*Cornus amomum*) and the more complex compound, whorled leaves of elderberry (*Sambucus canadensis*), whose fruit is used to make elderberry wine. (Don't sample the fruit though — the raw berries are reported to be toxic!) The marsh forms a buffer around three sides of the small pond along Huron River Drive near the Barton Dam parking lot. The edges of the pond attract green herons in the spring, and turtles can often be seen basking in the sun on rocks or logs. The bright yellow flowers and round green leaves of marsh marigold (*Caltha palustris*) on the shore of the pond are some of the first spring colors seen by passing motorists or bicyclists.

Other Wildlife

One of the few confirmed nesting spots for Turkey Vultures in Ann Arbor is in Bird Hills, where a pair of these large birds nests in a hollow old sugar maple. Hooded Warblers sometimes spend the summer on wooded slopes in Bird Hills. A pair of red fox have used a den within the park for many years.

Red fox prefer open sites with reliable cover nearby, frequenting forest/field edges and fencelines. During the breed–ing season (spring), red fox form strong male–female bonds and stay close to their underground dens.

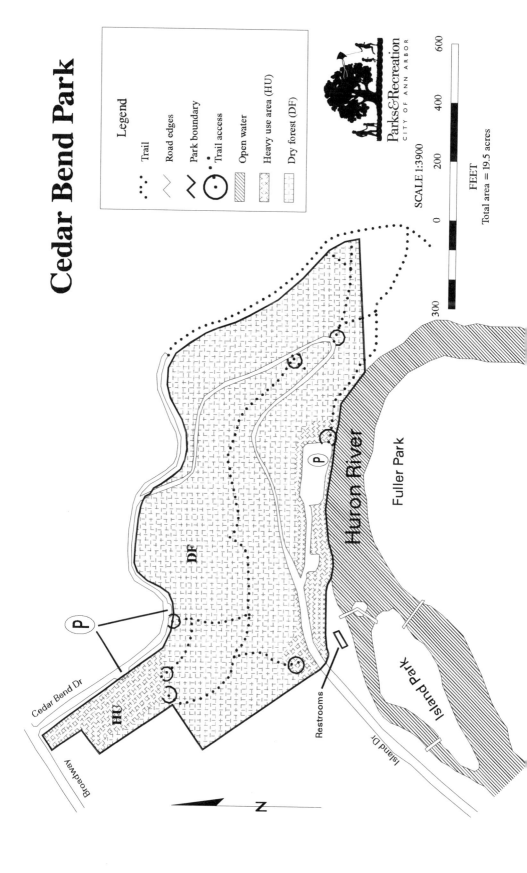

CEDAR BEND NATURE AREA

Parking	1) main lot Island Drive, 2) pull-offs on Cedar Bend Drive
Trail Access	at parking areas and along Island Drive
Restrooms	available at nearby Island Park
Picnic Areas	available at nearby Island Park
River Access	a few nearly inaccessible fishing spots along the bank
Comments	19.5 acres; bridge to historical shelter (at Island Park); scenic overlooks

General Information

Cedar Bend Nature Area is located on the steep bank of the river where the waterway makes a sharp bend back to the south. It is the high, forested slope seen across the river from Fuller Park. Because of the steepness of the site, trails in the park are minimal, although there are several unimproved paths along the slope and several trails along the river bank used for fishing. These trails are very eroded and slippery; please use caution if you choose to walk them. The easiest way to enjoy the interior of the park is along the shoulder of the gravel one-way road which winds its way downhill through the center of the park from Cedar Bend Drive to Island Drive.

History

Cedar Bend is one of the oldest parks in Ann Arbor. It was designed in the early 1900s by landscape architect O.C. Simonds, who also designed the Nichols Arboretum. Simonds was a pioneer in the art of designing landscapes to look natural. Cedar Bend was a showcase for his talents. In a report to the Ann Arbor Parks Commission in 1905, O.C. Simonds wrote of Cedar Bend Park, "...one gets beautiful views of the city and valley of the Huron. The river banks and portions of the hillside are covered with attractive native trees and shrubs. Every city should try to secure for posterity an attractive native woodland. It is not so important to develop the park by introducing carefully kept lawns and flower beds, but it is important to retain the native growth."

Cedar Bend has spectacular views of the Huron River valley. At one time Cedar Bend Drive continued around the northeast edge of the park and is said to have been the old "lovers' lane" because of the romantic views over the river. The old route is still a scenic path, although the growth of vegetation has partially blocked the view.

Located off the turnout at the top of the hill are the remnants of an old gazebo from the 1900s which used to offer picturesque views of the city.

Natural Communities

The entire wooded portion of the park is **dry forest**, although not as open as it once was, as indicated by several large, spreading trees now being crowded by younger competitors. From the heavy use area near Cedar Bend Drive, several large open–grown apple trees provide an excellent gateway to the interior of the forest. In the 1994–1998 plant inventory, 194 species of plants were recorded here, 143 of them native.

The woodchip path leading out of this area marks the location of extensive ecological restoration work begun in fall 1995. The steep slope below the road was cleared of most invasive woody plants such as common buckthorn (*Rhamnus cathartica*), honeysuckle (*Lonicera*), and the especially–prevalent tree–of–heaven (*Ailanthus altissima*), often known as "the tree of Brooklyn" after a poem by a similar name. After the removals, the site was replanted with native oak (*Quercus*) and serviceberry (*Amelanchier*) trees. Look for the young seedlings along the path.

In early spring, look for the bright yellow flowers of the uncommon early buttercup (*Ranunculus fascicularis*) growing in this dry forest. In summer you may see the large yellow urn–like flowers of smooth false foxglove (*Aureolaria flava*) strung along its robust leafy stalk. This plant grows in association with oak trees, on which it is said to be semi–parasitic. Along the slopes are the well–named hillside blueberry (*Vaccinium pallidum*), a small shrub with berries too small and sparse to inspire jam–making.

The main east-west trail offers an excellent cross–section of the park, transecting the dry forest dominated by oak (*Quercus*) and hickory (*Carya*). As the trail crosses a ravine in the center of the park, the temperature drops slightly. This area supports a slightly richer array of plants, cascading down the slope along the water course. Here you can find skunk–cabbage (*Symplocarpus foetidus*), Jack–in–the–pulpit (*Arisaema triphyllum*), and alternate–leaf dogwood (*Cornus alternifolia*), the one dogwood shrub which does not have twigs arranged on opposite sides of the branch.

The trail closest to the river is a slight contrast to the rest of the woods. It is quite open, looking more like this entire forest would have looked historically. Look for the attractive sinewy trunks of hornbeam (*Carpinus caroliniana*), also called musclewood or blue beech.

Other Wildlife

Recently Cooper's Hawks have nested within the upper portion of the park.

The dry forest of Cedar Bend Park contains smooth false foxglove (Aureolaria flava), with its yellow, urn–like flowers. This flower is often found in association with oak trees, on which it is said to be semi–parasitic.

Foster (Barton Park)

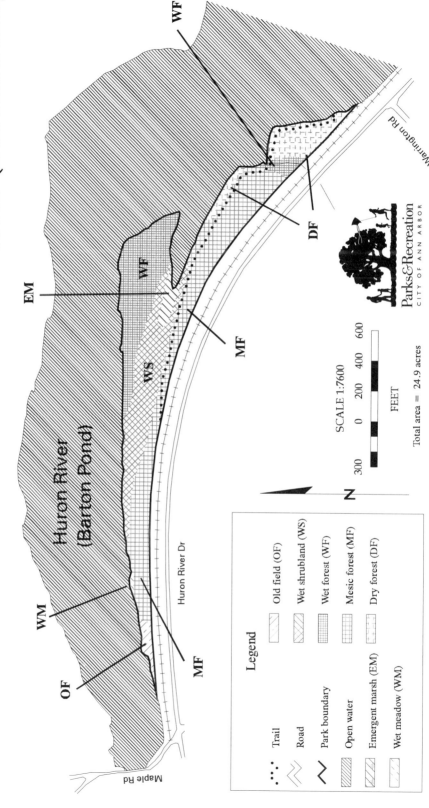

Huron River (Barton Pond)

Maple Rd

Huron River Dr

Warrington Rd

WF

EM

WM

OF

MF

WS

WF

MF

DF

N

SCALE 1:7600

FEET

300   0   200   400   600

Total area = 24.9 acres

Parks&Recreation
CITY OF ANN ARBOR

Legend

- ⋯ Trail
- Road
- Park boundary
- Open water
- Emergent marsh (EM)
- Wet meadow (WM)

- Old field (OF)
- Wet shrubland (WS)
- Wet forest (WF)
- Mesic forest (MF)
- Dry forest (DF)

FOSTER (BARTON NATURE AREA)

Parking	none, accessible only from the water
Trail Access	several canoe landings along the shore
Restrooms	none
Picnic Areas	none
River Access	numerous fishing sites along the shore of Barton Pond
Comments	24.9 acres

General Information
Foster is officially an extension of Barton Nature Area to the southeast, but due to its distinct location and ecological qualities this guide treats it as a separate park. Foster is a thin strip of land located between Barton Pond and the Conrail track near Huron River Drive, northwest of Barton Dam. Because the railroad is privately owned, the only way to access Foster legally is by boat. East of the stream there are several informal fishing trails that criss–cross the site. West of the stream is a narrow trail that runs the length of the park. Short spur trails provide fishing access along the river.

Unlike many of the Huron River park areas, Foster is a completely undeveloped piece of park land. Sadly, Foster has suffered tremendous abuse over the years from picnickers and partyers who have left piles of garbage, trampled vegetation, and cut trees.

History
Foster is the old name of a small cluster of houses where the train used to stop. This community is gone now, but the name remains in Foster Road, Foster bridge, and the name of this natural area.

The park property is in the former path of the railroad tracks, as indicated by the straight, flat ditch cutting across the eastern end of the park. Whether this was an earlier route of the main tracks or just a short spur section is unknown.

Foster is reputed to have been a study site for the renowned botanist Dr. Henry Gleason during his tenure at the University of Michigan in the 1930s. Dr. Gleason is well known as a coauthor with Dr. Arthur Cronquist of the *Manual of Vascular Plants of Northeastern United States and Adjacent Canada*.

Natural Communities
Foster contains at least seven natural communities: dry forest, mesic forest, wet forest, wet shrubland, emergent marsh, wet meadow, and old field. In the 1994–

1998 plant inventory, 230 species of plants were recorded here, 179 of them native.

The **dry forest** in the park is of the type also referred to as oak savanna. This is a unique mixture of widely spaced trees with a mixture of wildflowers and grasses beneath them. Many of the species in oak savannas may also be found in prairies, giving rise to several other local names for this site: Foster prairie, Foster woodland prairie, and Foster savanna. This is one of the highest quality oak savannas in our park system.

Fire is a vital component of the oak savanna community, and in its absence, the site has become somewhat overgrown with shrubs. Dead shrubs and small trees are evidence of prescribed burns in by NAP staff. The fires have been allowed to burn onto the railroad right–of–way, thanks to a cooperative agreement with Conrail. Native plants such as prairie dock (*Silphium terebinthinaceum*), tall coreopsis (*Coreopsis tripteris*), and golden Alexanders (*Zizia aurea*) have all spread after the fire.

Look for the large, bright green, oblong leaves of prairie dock (Silphium terebinthinaceum) in the dry forest of Foster Park.

West of the stream, the trail marks the boundary between dry forest along the river and **mesic forest** further inland on the higher ground. While these woods are thick with shrubs, a walk to the crest of the hill offers an excellent view of the expanse of big bluestem grasses (*Andropogon gerardii*) growing along the railroad tracks.

If you climb back into your canoe and paddle upstream a short distance you will see the area of **wet forest** on a peninsula jutting out into Barton Pond. Deer tracks sunk deep into the soil are evidence of ground made soft by frequent flooding. Many plant species are unable to tolerate such flooding; this has helped maintain the open nature of this community.

Older maps show the wet forest as an island in Barton Pond, but the river has been busy reconnecting it to the shore. Where this process is well underway, you'll find a **wet shrubland** community, thick with sandbar willow (*Salix exigua*) and dogwood (*Cornus*). Where the process is just getting started, at the back of the bay, you'll find this community blending into **emergent marsh** dominated by cat–tails (*Typha*).

Farther up river, along the shore, lies a small **wet meadow**. This open, grassy area hosts star grass (*Hypoxis hirsuta*), a slender yellow–flowered plant that is actually not a grass, and shrubby cinquefoil (*Potentilla fruticosa*), a small yellow–flowered shrub often used as an ornamental.

Finally, at the western tip of Foster, you'll find a tiny patch of **old field** growing on a site with disturbed soils close to the railway.

Other Wildlife

In *The Birds of Washtenaw County, Michigan*, Barton Pond is identified as a "hot spot" during the month of April. The book states, "there is probably no better time and place in the county for finding Common Loon, Horned Grebe, and Red-breasted Merganser." Many other spring migrants can be seen in the forests along the water's edge. While there, listen for the deep call of bullfrogs which have been heard near Foster's emergent marsh. Also look for chorus frogs and spring peepers in this park.

*Common Loons are spring migrants
stopping over at Barton Pond.*

Furstenberg Park

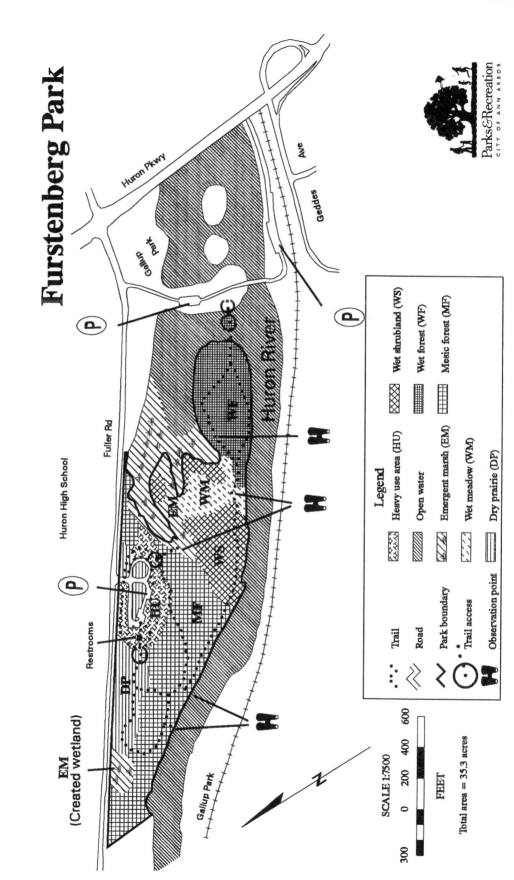

Legend

Trail	Heavy use area (HU)	Wet shrubland (WS)
Road	Open water	Wet forest (WF)
Park boundary	Emergent marsh (EM)	Mesic forest (MF)
Trail access	Wet meadow (WM)	
Observation point	Dry prairie (DP)	

SCALE 1:7500

FEET

300   0   200   400   600

Total area = 35.3 acres

Parks&Recreation
CITY OF ANN ARBOR

FURSTENBERG NATURE AREA

Parking	1) main lot on Fuller Road, 2) Gallup Park lots
Trail Access	parking areas
Restrooms	near main parking lot
Picnic Areas	scattered tables near main parking lot
River Access	two fishing platforms and a canoe launch along river
Comments	35.3 acres; interpretive panels and brochure, boardwalk and overlooks in wetlands

General Information
Furstenberg Nature Area is located on Fuller Road, across the street from Huron High School, and north and west of Gallup Park. Several paved or gravel loops begin and end near the main parking lot, and an extensive boardwalk connects to Gallup Park. The trails allow hikers to see a wide variety of natural communities. Interpretive brochures are available at several points in the park or from the Parks and Recreation Department office. Because the park is intended for quiet nature study, bicycles are not allowed on trails in Furstenberg Nature Area.

History
Although the City purchased the land for Furstenberg Park in the 1970s, serious work on the park did not begin until the early 1990s after extensive public input. Old junked cars and piles of rubble were cleared from the site, and trails and parking lots were laid out. Ecological restoration (described later) was begun in 1992. In 1995 a new wetland was created, and in 1996 the boardwalk, bridges from Gallup Park, interpretive displays, and restrooms were completed.

Natural Communities
One of the reasons that Furstenberg Nature Area has received so much attention in recent years is that it is such an ecologically rich site. In the 1994–1998 plant inventory, 337 species of plants were recorded here, 258 of them native. These numbers are the highest of any site inventoried, despite the relatively small size of the park. The natural communities found at Furstenberg include: dry prairie, mesic forest, wet shrubland, wet meadow, emergent marsh, and wet forest.

As you leave the parking lot and walk past the restrooms, you'll encounter the **dry prairie**. Unlike all other dry prairies in this guide, parts of this one were planted, beginning in 1992, but there are also dry prairie remnants near the edge of the forest. These remnants serve as a model and a seed source for our planted prairies. For an introduction to many of the following prairie species, stop by the exhibition planting near the park entrance before venturing out into the prairie. This patch was planted in 1995 as a way to help visitors identify the many wildflowers they will encounter in the park.

In the dry prairie, you can see a variety of showy yellow flowers: smooth false foxglove (*Aureolaria flava*), black–eyed Susan (*Rudbeckia hirta*), and prairie coneflower (*Ratibida pinnata*). Another tall yellow flower standing sentinel over the landscape is prairie dock (*Silphium terebinthinaceum*), most easily recognized by its platter–sized sandpapery leaves. Dotting the prairie, look for the orange flowers of butterfly–weed (*Asclepias tuberosa*) which, like other milkweeds, are food plants for the caterpillar stage of the Monarch butterfly.

A wide array of other butterflies provide additional color to the dry prairie throughout the spring, summer, and fall. These include the American and Bronze Copper and the Wild Indigo Duskywing.

Unfortunately, in the first few years of the prairie restoration, these native colors have been obscured by the pink thistle–like flowers of spotted knapweed (*Centaurea maculosa*), and the tiny but abundant flowers of white sweet–clover (*Melilotus alba*) and yellow sweet-clover (*Melilotus officinalis*). These persistent invasives have been the target of endless hours of pulling by NAP staff and volunteers. Slowly, their grip on the prairie is beginning to loosen. Periodic controlled burning is helping to control the invasives in this community.

At the edge of the dry prairie, where trees become more common, the landscape merges into **mesic forest**, which comprises most of the wooded sections of the park. If you look closely as you stroll the paths, you may still be able to see the cut stumps of the dense stand of buckthorn (*Rhamnus*) and honeysuckle (*Lonicera*) which were cut by staff and volunteers in the mid–1990s. It is hard to imagine the amount of work put into this restoration effort, or the dramatic change it has brought to the mesic forest. Where once there was dark shade and few wildflowers, there is now dappled sunlight and a sea of diverse flowers which have sprung back. Periodic fires set by trained staff provide a much less labor–intensive method of controlling these shrubs.

White trout lilies (Erythronium albidum) cover the forest floor in the mesic forest of Furstenberg. The leaves are spotted, resembling the coloring of a trout.

In spring, enjoy the carpet of white trout lilies

(*Erythronium albidum*) and Jack–in–the–pulpit (*Arisaema triphyllum*). Look carefully and you may also find the bright yellow flowers of early buttercup (*Ranunculus fascicularis*). This is also a good time and place to compare Solomon's seal (*Polygonatum biflorum*) and false Solomon's seal (*Smilacina racemosa*). The former has delicate drooping white flowers along the entire length of the stem, while the latter has a single large cluster of flowers borne only at the end of the plant.

Solomon's seal (*Polygonatum biflorum*) has white flowers along its entire stem.

Later in the summer, you can see nodding wild onion (*Allium cernuum*), pokeweed (*Phytolacca americana*), and late figwort (*Scrophularia marilandica*), a tall plant whose leathery, purplish–brown flower clusters are five feet off the ground.

As you begin strolling the boardwalk, you leave the tall trees of the mesic forest behind and enter a **wet shrubland** community, also called a "shrub carr" in the interpretive brochure for the park. Native shrubs such as red–osier dogwood (*Cornus stolonifera*), ninebark (*Physocarpus opulifolius*), and elderberry (*Sambucus canadensis*) are present here, but often obscured by the non–native glossy buckthorn (*Rhamnus frangula*) which is taking over the site.

The boardwalk has allowed extra sunlight to get through the dense shrub canopy, and you can see several colorful flowers along the trail in the summer: Michigan lily (*Lilium michiganense*), golden ragwort (*Senecio aureus*), and fringed loosestrife (*Lysimachia ciliata*). The latter is a native loosestrife with yellow flowers and is not related to purple loosestrife (*Lythrum salicaria*), the widespread wetland invader.

As you continue southeast along the boardwalk, the shrubs give way to non–woody vegetation, marking your entrance into the **wet meadow** community. The dominant ground cover here is the grasslike tussock sedge (*Carex stricta*) so–named for its mounded shape which keeps most of the plant above the several inches of water which are often present. Along the boardwalk, you may also see swamp betony (*Pedicularis lanceolata*), turtlehead (*Chelone glabra*), boneset (*Eupatorium perfoliatum*), sneezeweed (*Helenium autumnale*), or the tall hollow stem of angelica (*Angelica atropurpurea*) with its large leaves. Look for the magestic smooth swamp aster (*Aster firmus*), and the translucent stems of the spotted touch–me–not (*Impatiens capensis*). This is one of the spots where the Baltimore butterfly can be common.

In the middle of the boardwalk, enjoy the view of the backwater "lagoon" area from the observation deck toward Huron High School. You may see some of the charred remains of willow (*Salix*) or dogwood (*Cornus*) shrubs which were killed when staff burned this area in 1995 or a subsequent year.

You can also see cat–tails (*Typha*) but these mark the edge of the **emergent marsh** community. According to 1947 aerial photos, this marsh and the area where the observation deck is now located were once in the middle of the Huron River! But sedimentation by the river closed off this channel, just as it is now filling in this marsh. The cattails currently growing at the very edge of the open water at the back of the bay will eventually find themselves in the middle of the marsh as more cat–tails appear to march out into the shallow water. This is a natural process, although one that is accelerated by urban runoff into the river.

Look for the arrow–shaped leaves of arrow–arum (*Peltandra virginica*) and the deadly poisonous water hemlock (*Cicuta bulbifera*) also growing in the marsh.

There's another small patch of emergent marsh to be found at the extreme northwest corner of Furstenberg Park. It grows at the margins of a wetland mitigation —a wetland created to "mitigate" or lessen the impact of destroying an existing wetland. In this case, the wetland was lost in 1995 when Fuller Road was re–routed just north of the park. As part of the arrangement, a new wetland was built here, in a shrubland that was mostly overgrown with invasives. Although this wetland in no way *replaces* the wetland that was lost, it does serve some of the same ecological functions, such as collecting and cleaning runoff from Fuller Road before it flows into the Huron River.

As you leave the boardwalk and the emergent marsh, you enter the final community type in Furstenberg Park, the **wet forest**. This flat, low–lying area is only a few feet, at most, above the river. In the 1947 aerial photos, this was an island in the river, but its history before that is unclear. What is unusual about this wet forest is how heavily invaded it is by non-native plants. One possible explanation is that this area has been heavily disturbed in the past, possibly by having dredge material from the river dumped onto it. Whatever the reason, it doesn't matter to the birds, who seem to find it a popular stopping point during migrations.

There are, however, some native wildflowers to look for here. Green dragon (*Arisaema dracontium*), Michigan lily (*Lilium michiganense*), and richweed (*Collinsonia canadensis*) may all be spotted easily from the trail.

An uncommon butterfly, the Harvester, can be found in the wet forest and in other areas of Furstenberg where the shoreline is dominated by the invasive black alder tree (*Alnus glutinosa*).

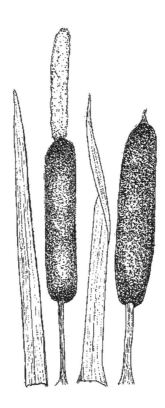

Broad–leaved cat–tails (Typha latifolia) are a native species but are still some–times considered to be invasive. In the emergent marshes of Furstenberg and other wetlands where excess nutrients are present, cat–tails aggressively take over, often to the exclusion of other native plants.

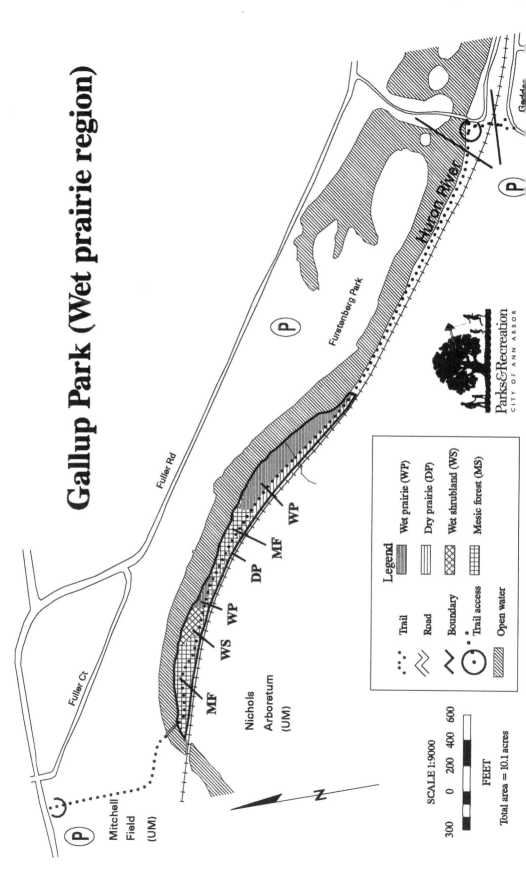

Gallup Park (Wet prairie region)

Fuller Ct

Fuller Rd

Mitchell Field (UM)

Nichols Arboretum (UM)

MF

WS WP

DP

MF

WP

MF

Furstenberg Park

Huron River

Geddes

N

SCALE 1:9000

300   0   200   400   600

FEET

Total area = 10.1 acres

Legend

Wet prairie (WP)

Dry prairie (DP)

Wet shrubland (WS)

Mesic forest (MS)

Trail

Road

Boundary

Trail access

Open water

Parks&Recreation
CITY OF ANN ARBOR

GALLUP PARK - WET PRAIRIE REGION

Parking	1) several lots in Gallup Park, 2) Mitchell Field on weekends and after 6:00 PM
Trail Access	paved bike trail from either parking area
Restrooms	1 mile away at Gallup Park canoe livery
Picnic Areas	1 mile away near Gallup Park canoe livery
River Access	nearby in other parts of Gallup Park
Comments	10.1 acres

General Information

Although Gallup Park is one of the city's most popular public areas, the most significant natural part of the park— the wet prairie— is not well known to many people. Gallup Park itself is located on either side of the Huron River near Huron Parkway. The wet prairie region of the park is along the southwest side of the river, directly between Furstenberg Nature Area and the Nichols Arboretum. It is stretched out between the river and the Conrail tracks, and is bisected from end to end by the much-traveled Gallup Park paved trail. This is an 8–foot wide, paved trail which runs for three miles along the river, from the University of Michigan's Mitchell Field on the west to Parker Mill on the east. It is shared by hundreds of bikers, runners, hikers, in–line skaters, and strollers every day. Because there are no other trails through the wet prairie region, the site should be enjoyed from this paved trail.

To reach the wet prairie from the Gallup Park entrance, walk through the first parking lot and over the one–lane wooden bridge over the river. At the south end of the bridge, turn right (west) and follow the bike trail for approximately one mile to the west.

History

The ecological significance of this area of the park has been known to the city since at least the late 1960s, when the late Paul W. Thompson of the Michigan Natural Areas Council lobbied to have it protected as a natural area. His 1970 article in the *Michigan Academician* helped document the richness of the site and provided a detailed map of the natural communities present at that time. This map and article have been extremely valuable to the NAP staff's recent restoration efforts. The article also helped establish the regional significance of the Gallup Park Wet Prairie, which has also been referred to as the Ann Arbor Wet Prairie, or simply the Ann Arbor Prairie. In a 1970s listing of natural communities around the state, the Michigan Chapter of The Nature Conservancy listed this site as one of only three good examples of wet prairie in the state.

Natural Communities

In addition to the wet prairie, the site contains three other natural communities: dry prairie, mesic forest, and wet shrubland. Despite its small size, this area is extremely diverse, containing 246 native plant species, two of which are listed as threatened or special concern in Michigan. In the 1994–1998 plant inventory, a total of 330 species of plants were recorded here.

As you enter the site walking northwest on the paved trail, the first natural community you come to is the **wet prairie**, the only one of its kind in Ann Arbor. It is between the river and the trail, from which you are high enough to look over most of its expanse. In spring, look for the bright yellow flowers and green leaves of marsh marigold (*Caltha palustris*), or the bluish–green shoots of tussock sedge (*Carex stricta*) rising out of the soggy ground. More careful scanning may reveal the uncommon tiny yellow flower of star

Turtlehead (Chelone glabra) is another plant growing in the Gallup wet prairie. Look for the distinctive white flowers that look like turtles' heads.

grass (*Hypoxis hirsuta*), or the single stalk of small flowers on swamp saxifrage (*Saxifraga pensylvanica*), which rises from a basal rosette of soft leaves. You may also see the rare sweetgrass (*Hierochloe odorata*), named for its sweet vanilla–like smell. It is the very earliest grass to bloom in Ann Arbor, and was used by Native Americans for basket-making. Later in the season, look for the tall flower stalks of prairie dock (*Silphium terebinthinaceum*), the white spires of culver's root (*Veronicastrum virginicum*), rising above leaves whorled at intervals along the stem, or the vibrant purple flowers of Missouri ironweed (*Vernonia missurica*). The wet prairie is one of the few sites along the corridor where the tall, rough leaves and coarse stalk of flowers of cordgrass (*Spartina pectinata*) can be seen.

The wet prairie has changed somewhat since 1970 when Paul Thompson drew his detailed map of the site. Purple loosestrife (*Lythrum salicaria*) has invaded and stands of cat–tails (*Typha*) have grown in size. Both the native species (*Typha latifolia*) and the exotic (*Typha angustifolia*) are present here and are an indicator of excess nutrients reaching the site, allowing them to flourish. Their aggressive advancement into the site is at the expense of other native species.

Another invasive tree has entered the wet prairie, but is restricting itself to the immediate edge of the Huron River. It is the black alder (*Alnus glutinosa*), the same tree that lines much of the riverbank throughout Gallup Park. It is easily recognized by the small cone–like female flowers, called catkins, which persist through the winter. These trees, and any woody species, are detrimental to the wet prairie. Because they draw moisture from the ground and release it to the atmosphere as they "breathe," they dry out the soil here. They also shade out the sun–loving wet prairie plants. The alder, however, does support populations of the orange and brown–colored Harvester butterfly. (See the section on *Butterflies* to find out the connection.) Baltimore butterflies are fairly common in the wet prairie as well.

The wet prairie was burned by trained parks staff on several occasions in the 1970s and 1980s. More recently, it was burned in the spring of 1994 and 1996. A Ph.D. candidate in biology at the University of Michigan, Dave Warners, has been researching the effects of these burns on the plants in this wet prairie. He found that in the first year following a fire, the burned portion of the wet prairie had a 90% increase in biomass (living plant matter) over the unburned portion! The burned portion also had a 83% increase in new plant shoots (seedlings). This stimulation of the native fire-adapted plants, coupled with the detrimental impacts to the woody species, is what makes prescribed fire such an important tool in ecological restoration.

Farther west along the trail but still on the river side, a **mesic forest** community has established itself in places. Of special note here is one of the largest stands of bladdernut (*Staphylea trifolia*) shrubs in Ann Arbor. These interesting shrubs depend on floods to disperse their seeds, which are enclosed in a hollow buoyant membrane. In the fall, the light green inflated fruits can easily be spotted in the understory (see illustration on page 1 of this guide).

Between the stands of mesic forest is a small patch of **wet shrubland**, dominated by the typical dense stand of willow (*Salix*), dogwood (*Cornus*), and ninebark (*Physocarpus opulifolius*). Of concern, however, is the encroaching stand of the invasive common reed or giant bulrush (*Phragmites australis*). This 9–foot tall grass with large plumes can often be seen in wet ditches along highways, where it forms solid stands.

On the opposite side of the trail, toward the railroad tracks, is a **dry prairie** community. Its higher elevation above the river and drier soils provide habitat for very different plants than the wet prairie. In late summer and fall, don't miss the beautiful big bluestem (*Andropogon gerardii*). This 6–foot tall grass is also called turkey foot because it produces large, 3–pronged seed stalks similar in shape to that fowl's foot. Less than half the size of big bluestem, little bluestem (*Schizachyrium scoparium*) also occurs here. Both species do indeed have a distinctive bluish hue to their stems in the summer. A third prairie grass is also present here, Indian grass (*Sorghastrum nutans*), with its beautiful amber color and soft seed heads waving gently in the breeze.

Other Wildlife
In early spring, American toads can be heard in loud choruses along the river here. A small colony of Cliff Swallows nests under the wooden car/foot-bridge that crosses the Huron River. In late May you can watch the birds collecting mud along the riverbank for the construction of their nests.

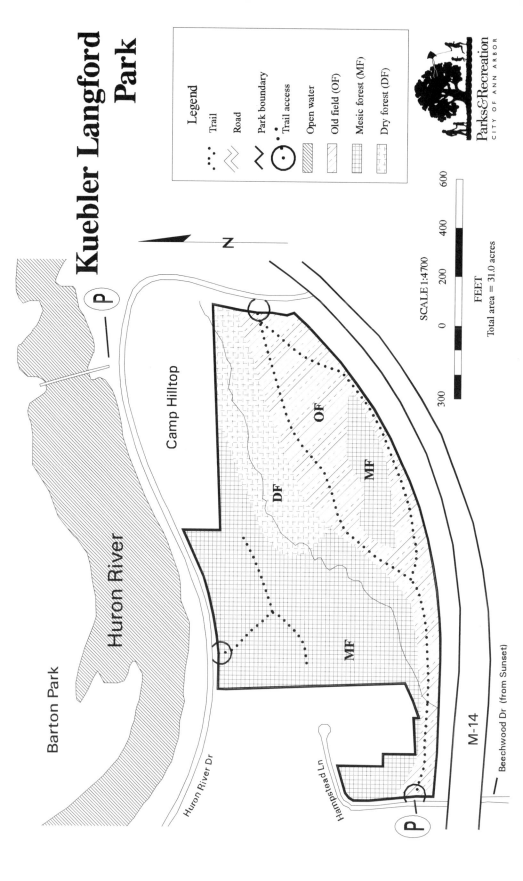

Kuebler Langford Park

Legend

· · · Trail
〉〉 Road
〉 Park boundary
⊙ Trail access
▨ Open water
▧ Old field (OF)
▦ Mesic forest (MF)
▤ Dry forest (DF)

Parks&Recreation
CITY OF ANN ARBOR

N

SCALE 1:4700

FEET

Total area = 31.0 acres

0   200   400   600

300

Barton Park

Huron River

Camp Hilltop

DF

OF

MF

MF

Huron River Dr

Hampstead Ln

M-14

Beechwood Dr (from Sunset)

P

P

KUEBLER LANGFORD NATURE AREA

Parking	1) pull-off on Beechwood Dr. from Sunset, 2) pull-off on Huron River Drive at Barton bridge from N. Main St.
Trail Access	parking areas; Huron River Drive
Restrooms	none
Picnic Areas	none
River Access	none
Comments	31 acres

General Information
Kuebler Langford Nature Area is located immediately north of M–14, where it is tucked inside the sharp curve near Huron River Drive. The property is bisected from northeast to southwest by a steep, wooded ravine. The most accessible trail is a wide, mowed grass path running between Beechwood Drive and Huron River Drive on the southeast side of the ravine, close to the highway. There is, however, a short spur trail on the northwest side of the park accessible only from an unmarked trailhead on Huron River Drive. Please note that Camp Hilltop is private property bordering the park and should not be used as either a parking area or an access route.

History
The property was used as a staging area during the construction of M–14 in the late 1970s. Most of the topsoil was removed and heavy machinery was stored on site. Underground pipes were also recently installed in this area. These activities led to compaction of the soil and an increase in soil erosion. The steep erosion gullies on the site are the result. The open areas nearest the highway are very disturbed, though the rest of the park is in remarkably good ecological condition.

Natural Communities
The open (old field) and wooded (dry and mesic forest) natural communities in Kuebler Langford roughly follow the boundary between the disturbed and undisturbed areas. In the 1994–1998 plant inventory, 296 species of plants were recorded here, 215 of them native.

In the disturbed areas, an **old field** community has developed. Because there is so little topsoil remaining in this part of the park, the soil is quite impoverished. Interestingly, some of the species which established themselves are native dry prairie species which can tolerate dry, nutrient-poor soils. Here you can find round–headed bush–clover (*Lespedeza capitata*), whose dense heads of small white summer flowers are not as conspicuous as the dark brown globes they turn into later in the fall. These dried seed heads persist through the winter and are easily seen against the white snow. In the fall, also look for the bright yellow flower clusters of stiff goldenrod (*Solidago rigida*).

Kuebler Langford is the only park in Ann Arbor which contains four species of tick–trefoil (*Desmodium*), named for the seeds which cling, tick-like, to clothing. In the old–field, look for three species: showy tick–trefoil (*Desmodium canadense*), small–leaved tick–trefoil (*Desmodium marilandicum*), and panicled tick–trefoil (*Desmodium paniculatum*).

The fourth species, clustered–leaved tick–trefoil (*Desmodium glutinosum*), is found in the **mesic forest** community of the park. Although there is a small island of forest in the old field, the major extent of it is across the ravine. This area is similar to the nearby ravines in Bird Hills Nature Area, and contains many of the same species of plants and animals. Here you can find maiden hair fern (*Adiantum pedatum*), bishop's cap (*Mitella diphylla*), sharp–lobed hepatica (*Hepatica acutiloba*), and richweed (*Collinsonia canadensis*). There's even a groundwater seep on a steep slope with very mucky soil that supports skunk–cabbage (*Symplocarpus foetidus*). One of the easiest plants to spot is false spikenard (*Aralia racemosa*), a 6–foot tall plant with large divided leaves and a spectacular display of numerous small dark fruit clusters. Kuebler Langford has one plant species that is unique to this city park: the oak fern (*Gymnocarpium dryopteris*), a delicate fern with a wiry stalk found on the upper slopes of the ravine.

Separating the old field and mesic forest, the **dry forest** is probably the oldest and least disturbed of Kuebler Langford's natural communities. Large black and white oaks (*Quercus velutina* and *Quercus alba*) cast moderate shade suitable for understory plants such as witch–hazel (*Hamamelis virginiana*), a delicate shrub unique for its yellow flowers that bloom in fall and persist into winter. Stands of hazelnut shrubs (*Corylus americana*) dot the woods, their clonal shoots springing up from the shallow roots. The edge of the woods nearest the old field allows enough light to reach the ground to support bracken fern (*Pteridium aquilinum*) amongst a carpet of Pennsylvania sedge (*Carex pensylvanica*). On the ravine's slope, bloodroot (*Sanguinaria canadensis*) abounds.

Other wildlife
Blue–winged Warblers, otherwise uncommon in Ann Arbor, are sometimes found during the summer in Kuebler Langford. Deer are common visitors to the woods and open areas of this park.

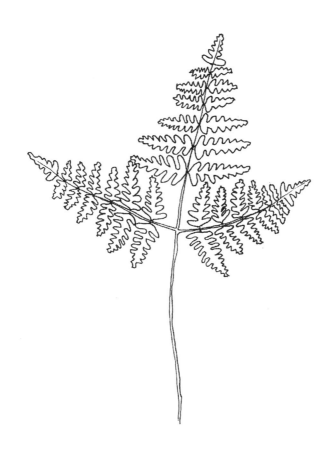

Oak fern (Gymnocarpium dryopteris) is not found in any other city park besides Kuebler Langford. Look for it on the upper slopes of the ravine.

Nichols Arboretum

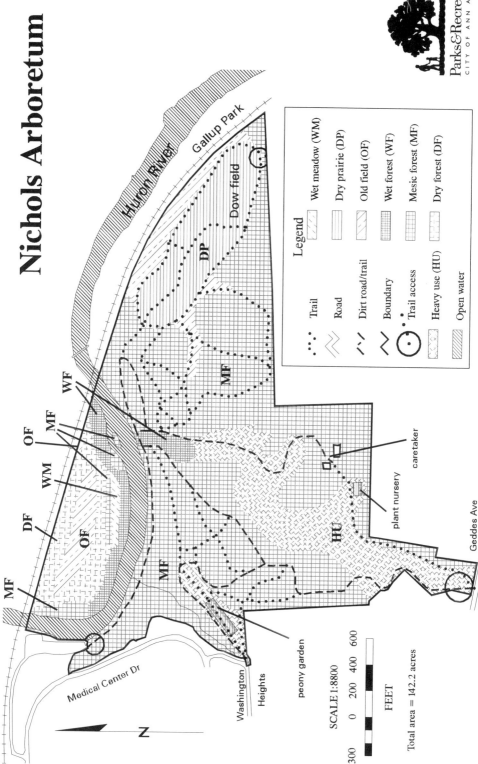

Huron River

Gallup Park

Medical Center Dr

Washington
Heights

peony garden

Dow field

DP

MF

WF

MF

OF

WM

DF

OF

MF

MF

plant nursery

caretaker

HU

Geddes Ave

N

SCALE 1:8800

300   0   200   400   600

FEET

Total area = 142.2 acres

Legend

	Trail		Wet meadow (WM)
	Road		Dry prairie (DP)
	Dirt road/trail		Old field (OF)
	Boundary		Wet forest (WF)
	Trail access		Mesic forest (MF)
	Heavy use (HU)		Dry forest (DF)
	Open water		

Parks&Recreation
CITY OF ANN ARBOR

NICHOLS ARBORETUM

Parking	UM Medical Center lower lot (Lot M29) on E. Medical Center Drive, except 6:00 AM to 6:00 PM Monday-Friday
Trail Access	parking areas; Geddes Avenue entrance, Washington Heights entrance
Restrooms	none currently (planning is underway for a new interpretive center)
Picnic Areas	no tables, many open mowed areas
River Access	several fishing spots along river bank
Comments	123 acres, plus 19.2 acres across the river; interpretive brochure

General Information

The Nichols Arboretum (locally known as the "Arb") is located between Geddes Road and the Huron River just east of the University of Michigan Hospital. The land in the Arb is owned by the University of Michigan and the City of Ann Arbor. The city owns approximately 25 acres near the west entrance. The UM School of Natural Resources and Environment, however, is responsible for managing all of the Arb property. The isolated, undeveloped piece of land across the river from the Arb (see the accompanying map), is also owned by the University but is not considered part of the Arb. It is included here because it, too, is a significant natural area along the Huron River corridor. This area is accessible only by boat.

All trails in the Arb are unpaved, but there are several hard–surface service drives indicated on the map. From the river to the Geddes Avenue entrance is a 185–foot rise in elevation— take your time if you're not in shape!

History

Nichols Arboretum was founded in 1907 with a 27.5–acre gift of land from Ester and Walter Nichols. In 1943, 36 acres of land were given to the University by Alex Dow, what is now known as Dow Prairie.

The Arb was designed by O.C. Simonds, the famous landscape architect who also designed Cedar Bend Nature Area and founded the UM Department of Landscape Design. Simonds used a variety of plant materials to emphasize the diverse landscape and physical features of the site. To increase the variety of plants available for study, many exotic species were brought into the Arb from all over the world.

Trees and shrubs are still being planted in the Arboretum. Many of them are marked with scientific name, common name, and accession number. The accession number consists of the year of planting and the plant number. The color of the labels indicates whether or not the plant is native to Michigan or to North

America: green is native to Michigan, red is native to North America, and plain silver metal is exotic to North America. For more information, consult the interpretive brochure available at the Geddes Avenue or Washington Heights entrances, or from the Arb office at the University of Michigan— School of Natural Resources and Environment.

Natural Communities

Although NAP staff have not conducted extensive plant or animal surveys at the Arb or the other University property across river, there appear to be at least six natural communities found there: mesic forest, dry prairie, wet forest, dry forest, wet meadow, and old field.

Most of the Arb property is classified as **mesic forest**. This is a good place to learn to recognize red oak (*Quercus rubra*), white ash (*Fraxinus americana*), tuliptree (*Liriodendron tulipifera*) and black walnut (*Juglans nigra*), common mesic forest trees which are identified in the Arb's interpretive brochure.

One of the largest examples of a **dry prairie restoration** in our area can be found in Dow Prairie, at the eastern end of the Arb. One–hundred sixteen (116) different plant species have been recorded here, 81 of them native. Here, hikers can walk through a large field of big bluestem (*Andropogon gerardii*), little bluestem (*Schizachyrium scoparium*), Indian grass (*Sorghastrum nutans*) and goldenrod (*Solidago*). The prairie is burned regularly by university staff who are researching the effectiveness of both spring and fall burns with varying frequency. This is the oldest example of dry prairie restoration on public land in the city.

Where the land levels out at the bottom of the ravine in the middle of the Arb, is an example of a **wet forest** community. This is also the forest type which borders the far side of the river. Here you can find the inflated seed capsules of bladdernut (*Staphylea trifolia*), and the spherical fruit of buttonbush (*Cephalanthus occidentalis*), an attractive shrub often found growing in flooded areas. Some of the showy wildflowers present here include great blue lobelia (*Lobelia siphilitica*), spotted touch–me–not (*Impatiens capensis*), and fringed loosestrife (*Lysimachia ciliata*).

Across the river from the Arb is a patch of **dry forest**. In addition to the usual invasive shrubs which encroach into dry forests, this area also supports a stand of the non–native Scots pine (*Pinus sylvestris*), easily identified from a distance by its orange flaking bark.

On this same side of the river are two open types of natural communities: a **wet meadow**, which contains tussock sedge (*Carex stricta*), turtlehead (*Chelone glabra*), and Joe–pye weed (*Eupatorium maculatum*), and an **old field** which is just slightly upslope on drier ground. Patches of native prairie plants growing here indicate that the site was probably dry prairie before being used for agriculture. Some of these indicators include: showy goldenrod (*Solidago speciosa*), butterfly–weed (*Asclepias tuberosa*) prairie dock (*Silphium terebinthinaceum*), flowering spurge (*Euphorbia corollata*), big bluestem (*Andropogon gerardii*), and little bluestem (*Schizachyrium scoparium*).

Other Wildlife

The magnitude of the spring and fall bird migrations is legendary in Nichols Arboretum. Up to 35 species of warblers are seen here annually, with many records of rare or vagrant birds which have appeared for a day or two. For years a pair of Cooper's Hawks have nested in the Arb, and Broad–winged Hawks have also nested here on several occasions.

The Carolina Wren , sometimes spotted in the Arb, is an uncommon permanent resident of the Ann Arbor area. These small brown birds with cocked tails have been making a comeback after the Michigan population crashed during several cold, snowy winters in the late 1970s.

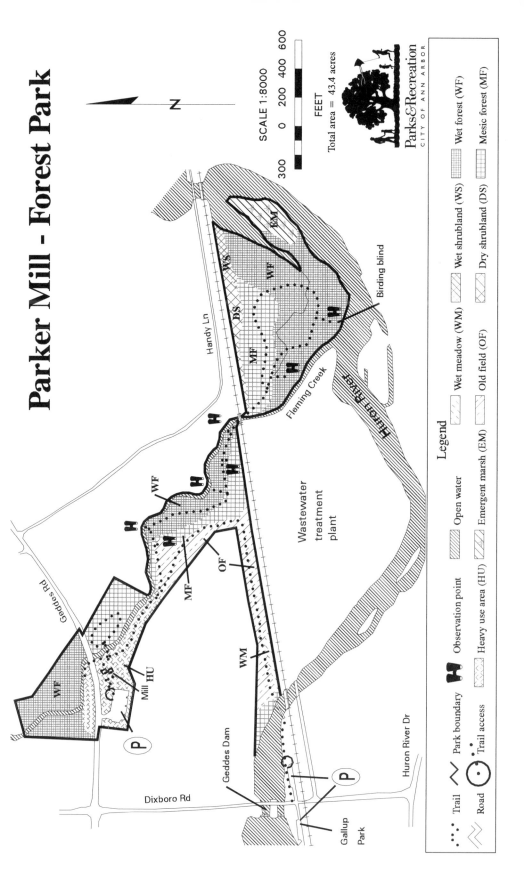

Parker Mill - Forest Park

SCALE 1:8000

FEET

Total area = 43.4 acres

300 0 200 400 600

N

Parks&Recreation
CITY OF ANN ARBOR

Legend

Trail

Road

Park boundary

Observation point

Trail access

Heavy use area (HU)

Open water

Emergent marsh (EM)

Wet meadow (WM)

Wet shrubland (WS)

Dry shrubland (DS)

Wet forest (WF)

Mesic forest (MF)

Old field (OF)

Geddes Rd

Handy Ln

Fleming Creek

Huron River

Wastewater treatment plant

Birding blind

Dixboro Rd

Geddes Dam

Huron River Dr

Gallup Park

Mill HU

WF

MF

OF

WM

WS

DS

MF

WF

EM

PARKER MILL - FOREST NATURE AREA

Parking	1) main lot on Geddes Road, 2) Dixboro Rd
Trail Access	parking lots
Restrooms	at the mill when open (call Washtenaw County Department of Parks and Recreation at 734-971-6337 for the schedule)
Picnic Areas	tables and grill by old mill buildings
River Access	numerous fishing sites along river bank
Comments	43.4 acres; old Parker mill; Hoyt G. Post Memorial Trail; wildlife blind; peat dome

General Information
Parker Mill is located between Geddes Road and the Huron River, east of Dixboro Road. Parker Mill is owned and managed by the Washtenaw County Department of Parks and Recreation. Combined with Parker Mill in this guide is the area south of the railroad tracks, Forest Nature Area, which the county leases from the City of Ann Arbor.

There is a wide paved trail and boardwalk which connects the main parking lot on Geddes with the lot on Dixboro. This trail connects to the east end of the Gallup Park bike trail, and is also open to bicycles. The Hoyt G. Post memorial trail branches off the main trail and follows Fleming Creek under the railroad bridge and down into Forest Park. This is a boardwalk intended as a quiet nature study trail and is not open for bicycling, in-line skating, or running.

History
The park is the old homestead of the Parker family, who ran a mill on the property. Some of the buildings are still standing next to the main parking lot. Next to the mill is a huge, open-grown bur oak (*Quercus macrocarpa*) which shades a log cabin. At one time, this cabin was the residence of the Parker family.

The county bought the property in the late 1970s and in 1996 completed an extensive renovation of the facilities, including construction of the hard-surface trails and boardwalks. The interpretive loop is named in honor of Hoyt G. Post, a naturalist who loved to walk along Fleming Creek and the Huron River.

Natural Communities
Parker Mill is an extremely diverse site, containing seven mapped natural communities: mesic forest, wet forest, emergent marsh, wet shrubland, dry shrubland, old field, and wet meadow. In the 1994–1998 plant inventory, 238 species of plants were recorded here, 177 of them native.

As you walk along Fleming Creek, the first area of forest you enter is a **mesic forest**. In the spring, visitors can enjoy the flowers of waterleaf (*Hydrophyllum virginianum*), yellow trout lily (*Erythronium americanum*), large-flowered trillium (*Trillium grandiflorum*), wild ginger (*Asarum canadense*), and May apple (*Podophyllum peltatum*). In the summer, stickseed (*Hackelia virginiana*) can be seen flowering here. The mesic forest in Parker Mill is the only location along the Huron River corridor where nodding trillium (*Trillium flexipes*) and Sprengel's sedge (*Carex sprengelii*) are known to occur.

At the beginning of the Post memorial boardwalk you enter an **old field** community. In addition to the European species typically found here, there are some interesting natives. You can find big bluestem (*Andropogon gerardii*), golden Alexanders (*Zizia aurea*), and bastard-toadflax (*Commandra umbellata*), an unassuming plant with small white flowers which is parasitic on the roots of other plants.

The umbrella-shaped leaves of May apple (Podophyllum peltatum)

A little further along the Post memorial boardwalk trail, you enter one of the most intact examples of **wet forest** found near Ann Arbor. Compared to other such forests in the city, the wet forest here, and especially in the Forest Nature Area, has large trees and is relatively free of invasive plants. One of the dominant trees found here is black maple (*Acer nigrum*), distinguished from sugar maple (*Acer saccharum*) by its leaves which are hairy on the underside veins and have drooping tips. Other uncommon trees found in this wet forest are northern hackberry (*Celtis occidentalis*), with distinct sharp ridges running up and down the trunk, and rock elm (*Ulmus thomasii*), a unique tree with thick corky bark and winged twigs.

One of the highlights of this wet forest is the peat dome near the mouth of Fleming Creek, which is explained nicely in the interpretive panel at that lookout. In early spring, look for the abundant skunk-cabbage (*Symplocarpus foetidus*) pushing up through the snow. Later in spring and summer, you can smell a crushed leaf of this plant to learn where it gets its common name. On the fringes of the peat dome are swamp saxifrage (*Saxifraga pensylvanica*) and ostrich fern (*Matteuccia struthiopteris*), a robust fern producing a circle of waist–high fronds.

Other flowers to enjoy in the wet forest include: spotted touch–me–not (*Impatiens capensis*), wood anemone (*Anemone quinquefolia*), green dragon (*Arisaema dracontium*), Jack–in–the–pulpit (*Arisaema triphyllum*), wild leek (*Allium tricoccum*), and hops (*Humulus japonicus*), a rough, barbed vine twining over shrubs and trees along the boardwalk. This is also one of the few places in Ann Arbor where the spiny–fruited bur cucumber (*Sicyos angulatus*) can be found.

This wet forest section of Parker Mill is a refuge for woodpeckers in winter. Alert birders commonly see Downy, Hairy, and Red–bellied Woodpeckers, plus a variety of less common winter birds.

Bird–watching along the river is enhanced by a wildlife blind along the boardwalk. The river is normally kept open year–round at this point by a combination of the flow from Fleming Creek and the warm water discharge from the city's wastewater treatment plant just upstream. This is an excellent spot to go in search of winter waterfowl. Uncommon sightings there include American Coot, Tundra Swan, and three kinds of Mergansers (Common, Hooded, and Red–breasted).

Past the boardwalk loop, jutting out into the Huron River at the far end of the Forest Park portion of Parker Mill, is an **emergent marsh** community. During spring floods, this area is often underwater and year–round is too thick with cat–tails (*Typha*) and purple loosestrife (*Lythrum salicaria*) to easily walk through.

A little further from the river, nearer the railroad track is a **wet shrubland** community, where elderberry bushes (*Sambucus canadensis*) can be found. Wildflowers present here include golden ragwort (*Senecio aureus*) and fringed loosestrife (*Lysimachia ciliata*).

Still further inland, the wet shrubland merges into a **dry shrubland** community on the higher ground near the railroad tracks. Here you can find gray dogwood (*Cornus foemina*), a native shrub which forms large dense stands. This site is near the edge of the Conrail right–of–way, where there is a nice strip of dry prairie. The big bluestem (*Andropogon gerardii*) forms a swath on both sides of the tracks, providing a prairie corridor all the way to the prairie remnant less than a mile to the east (on private property). If the invasive buckthorn (*Rhamnus*) and honeysuckle (*Lonicera*) shrubs were removed from the site, and the area treated with fire, the dry shrubland would likely develop into a dry prairie.

As you leave Parker Mill to the west, the last natural community you pass is a **wet meadow**, on the slope between the boardwalk and the railroad tracks. In the spring it is dominated by tussock sedge (*Carex stricta*), which is overshadowed by Joe–pye weed (*Eupatorium maculatum*) and angelica (*Angelica atropurpurea*) later in the summer. Also found in the wet meadow are marsh pea (*Lathyrus palustris*), and sweet grass (*Hierochloe odorata*). All of these can be easily viewed from the boardwalk; the thick, wet, decaying vegetation makes leaving the boardwalk difficult.

Other Wildlife
Recently, two butterfly species, the Hackberry and Tawny Emperor, have been found along the wooded trails of the park. Both species depend on hackberry trees for egg-laying and larval development. Eastern Screech Owls and Great Horned Owls nest in the park as well.

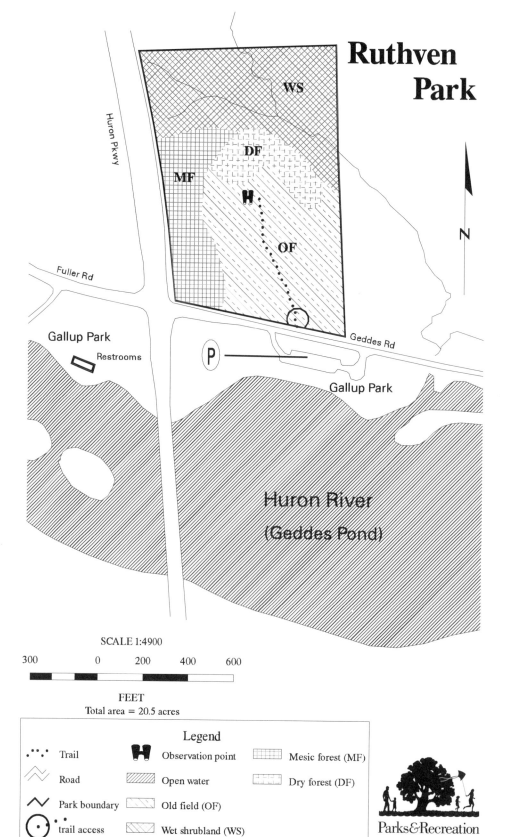

Ruthven Park

WS

DF

MF

H

OF

Huron Pkwy

Fuller Rd

Gallup Park

Restrooms

P

Geddes Rd

Gallup Park

Huron River

(Geddes Pond)

N

SCALE 1:4900

300   0   200   400   600

FEET
Total area = 20.5 acres

Legend

- · · · · Trail
- 〰 Road
- ⌢ Park boundary
- ⊙ trail access
- 🔭 Observation point
- ▨ Open water
- ▨ Old field (OF)
- ▨ Wet shrubland (WS)
- ▦ Mesic forest (MF)
- ▦ Dry forest (DF)

Parks&Recreation
CITY OF ANN ARBOR

RUTHVEN NATURE AREA

Parking	boat launch on Geddes Road in Gallup Park
Trail Access	Geddes Road, just east of Huron Parkway
Restrooms	none
Picnic Areas	none
River Access	none
Comments	20.5 acres; glacial kame

General Information

Ruthven Nature Area is located on the northeast corner of the Huron Parkway and Geddes Road intersection. There is currently no sign designating it as a park, and no obvious trailhead. To access the site, look for the unmarked footpath leading north from Geddes Road next to the Huron Parkway sign. The informal trail winds its way up to the top of the glacial kame where you can get a beautiful view of the Huron River valley, Geddes Pond and Gallup Park. (See the section on *Reading the Landscape* for a complete explanation of a kame.)

History

In addition to the geological clues about the site's glacial history, there is at least one biological clue of past life here. Just north of the park are some condominiums, easily visible from the top of the kame. During construction on the site, some boys were digging in the dirt and found what they believed to be an animal skull. They took it to the Natural History Museum at the University of Michigan to see if they were right. Sure enough, it *was* an animal skull, but a much older one than they had guessed. It was the partially–fossilized skull of an extinct peccary (a type of wild pig) which lived here about 14,000 years ago! The skull is still at the museum.

The property was purchased from Alexander and Florence Ruthven in 1966 to be used as a park and also to secure land needed for the future Huron Parkway, which was constructed several years later.

Natural Communities

Ruthven Park contains four natural communities: old field, dry forest, mesic forest, and wet shrubland. In the 1994–1998 plant inventory, 139 species of plants were recorded here, 88 of them native.

The trail takes you through only the **old field** community, dominated here by the non–native Canada bluegrass (*Poa compressa*), smooth brome (*Bromus inermis*) and Queen Anne's lace (*Daucus carota*), also called wild carrot. In early fall, the most conspicuous native plants are goldenrod (*Solidago*) and sumac (*Rhus*), whose leaves and erect seed clusters turn a beautiful crimson color. One of the native wildflowers found on this dry site is the very aromatic bee–balm (*Monarda fistulosa*), also called wild bergamot. If you miss the lavender flowers in the summer, you can still enjoy the minty smell of the seed head which persists through winter. You can also find butterfly–weed (*Asclepias tuberosa*), which is a type of milkweed, and dogbane (*Apocynum cannabinum*), which is a milkweed relative. Ironically, butterfly–weed does not have the milky latex of most other milkweeds, but dogbane does!

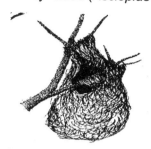

As you leave the trail and wander to the northeast, you quickly enter a disturbed **dry forest** community. You'll find that you have to fight your way through patches of autumn olive (*Eleagnus umbellata*), a plant once highly regarded for its wildlife value (mature plants have been know to produce 70

Oriole nests can often be spotted in the winter, hanging from tree branches

pounds of fruit in a season!), though now known to be highly invasive. Interspersed among the speckled stems of this shrub is one of the best examples of oak regneration in the parks. Here, scarlet oak (*Quercus coccinea*) seedlings and saplings are growing in large numbers in the shadows of red cedar (*Juniperus virginiana*) and older oaks. Note how deeply lobed the bristle–tipped leaves of the scarlet oak are in comparison to red or black oaks. On the north side of the kame, the low growing fern, ebony spleenwort (*Asplenium platyneuron*), can be found.

Further north, at the base of the kame, is a stream (known as the "North Campus Drain") running through a **wet shrubland** community. Along this stream, dense thickets of silky dogwood (*Cornus amomum*) bear metallic blue berries in the fall. You'll need to struggle through these shrubs if you want to find the bright blue monkey–flower (*Mimulus ringens*), known from only a few other sites in Ann Arbor. While you're there, look for elberberry (*Sambucus canadensis*), fringed loosestrife (*Lysimachia ciliata*), spotted touch–me–not (*Impatiens capensis*), and golden ragwort (*Senecio aureus*).

Here, you can hear frogs calling. In fact, it has the largest population of gray tree frogs of any park along the Huron River. Spring peepers and green frogs have also been heard here, although it is sometimes difficult to determine whether the frog calls are coming from within the park or from wetlands just north of the park.

On the west edge of the kame, along Huron Parkway, is a **mesic forest** community where you can find tall agrimony (*Agrimonia gryposepala*), horse–gentian (*Triosteum*), and wild geranium (*Geranium maculatum*). The hydrology of this site appears to have been altered by some type of disturbance in the past, perhaps associated with the construction of Huron Parkway. As a result, this area is wetter than it used to be, and may evolve into a different community over time.

Male Hooded
Merganser

Female Hooded
Merganser

Hooded mergansers are common migrants along the Huron River corridor. In other parts of the city, they nest in small wooded ponds.

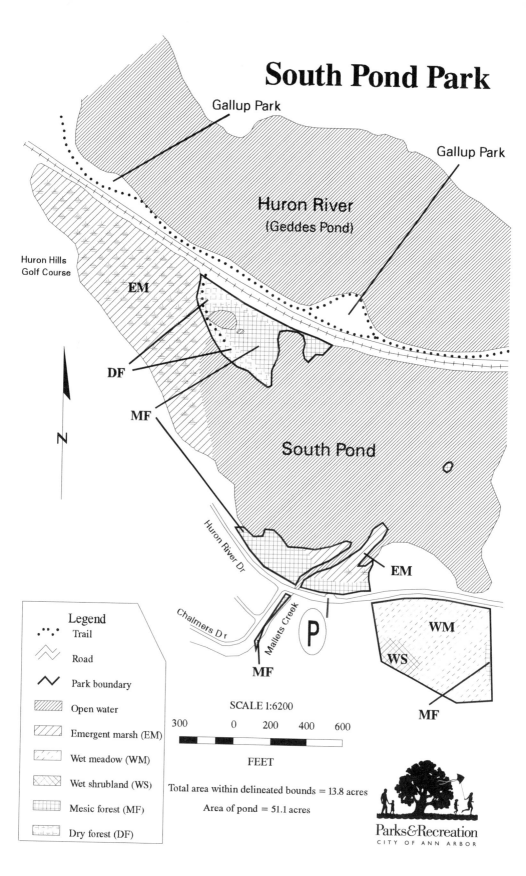

South Pond Park

Gallup Park

Gallup Park

Huron River
(Geddes Pond)

Huron Hills
Golf Course

EM

DF

MF

N

South Pond

Huron River Dr

Chalmers Dr

Mallets Creek

P

MF

EM

WM

WS

MF

Legend

- ····· Trail
- ∿ Road
- ⋀ Park boundary
- ▨ Open water
- ▨ Emergent marsh (EM)
- ▨ Wet meadow (WM)
- ▨ Wet shrubland (WS)
- ▤ Mesic forest (MF)
- ▦ Dry forest (DF)

SCALE 1:6200

300   0   200   400   600

FEET

Total area within delineated bounds = 13.8 acres

Area of pond = 51.1 acres

Parks&Recreation
CITY OF ANN ARBOR

SOUTH POND NATURE AREA

Parking	pull-off on Huron River Drive just east of Chalmers Road
Trail Access	roadside gravel parking area at shore of pond
Restrooms	none
Picnic Areas	none
River Access	several fishing sites along shore of pond
Comments	13.8 acres of park land, plus 51.1 acres of water surface area

General Information

The name South Pond refers to both a body of water and to the park which contains it. Both are located south of the river, just north of Huron River Drive and east of Huron Parkway and Huron Hills Golf Course. A recent 8–acre addition to the park is located south of Huron River Drive, away from the pond. The small pull–off on Huron River Drive is the only way to reach the pond. Here, a canoe can be launched at the mouth of Mallet's Creek. The only public access to the rest of the parkland on the north side of the pond is by boat. The Gallup Park bike trail runs near the park, but on the other side of the railroad tracks, which are private property owned by Conrail and should not be crossed without their permission. There are only a few informal footpaths in the park.

History

In 1819, the General Land Office of the federal government was in this part of the "Northwest Territory" laying out the grid of twenty townships that today make up Washtenaw County. Every township has 36 sections, each a square measuring one-mile per side. As the surveyors walked along these lines, establishing the grid we know so well today, they made careful notes about the character of the landscape. It so happens that one of these north-south section lines runs through the edge of what is now South Pond. But in 1819, the area which is now pond was described as "prairie." What happened since then?

First, a railroad line was constructed. Then a dam was built a short distance downriver, raising the water level and flooding the railroad. So the railroad bed was built higher. The effect was to create a barrier between the mouth of Mallets Creek and the river, basically damming the creek and creating an impoundment which became South Pond. Finally, when US–23 was constructed, Swift Run Creek was re–routed so it flowed into the eastern edge of South Pond rather than into the river. These actions had dramatic impacts on the site. In addition, the property appears to have been dredged at some point. The dredging, plus the railroad construction, combined with the rising river level, created a pond where there had once been only prairie.

The open water of the pond attracts a variety of waterfowl, and exotic mute swans grace the site. The swans were introduced by neighborhood residents in an attempt to control the Canada goose population. Bullfrogs and green frogs can also be heard. But nature seems determined that a pond should not persist here. The pond is filling in with sediment and becoming a marsh. Aquatic plants have been harvested, most recently in the summer of 1996, in an effort to keep the pond clear. The source of the weed problem comes from upstream. Two county drains flow into the pond, Mallets Creek and Swift Run Creek. Both creeks drain a highly urbanized landscape and carry with them a load of excess nutrients and sediments. With the continued build–up of these sediments and nutrients, South Pond may eventually be filled in and become a marsh.

The area south of Huron River Drive once received the drainage of Malletts Creek before it was re–routed to flow more directly into South Pond. The large knee–high mounds of tussock sedge *(Carex stricta)* here are evidence that the water level was once much higher.

Natural Communities
South Pond Nature Area currently contains five natural communities: emergent marsh, dry forest, mesic forest, wet shrubland, and wet meadow. An additional community exists on the adjcent privately–owned railroad right–of–way. In the 1994–1998 plant inventory, 286 species of plants were recorded here, 215 of them native.

The **emergent marsh** has already been mentioned. It is slowly growing out from the mouth of Mallet's Creek on sediment carried to the pond from the stream. Besides the ever–present cat–tails *(Typha)* and purple loosestrife *(Lythrum salicaria)*, look for the pink flowers of swamp milkweed *(Asclepias incarnata)*, the early spring yellow flowers of marsh marigold *(Caltha palustris)* and swamp buttercup *(Ranunculus hispidus)*. Two kinds of iris are also seen here: the native southern blue flag *(Iris virginica)* and the non–native yellow flag *(Iris pseudacorus)*. The leaves seen floating on the water surface are of water–lily *(Nymphaea odorata)*.

The peninsula jutting into the pond from the north makes the effort of getting there well–worth your time. The **dry forest** community is of the type we refer to as oak savanna, with open–grown black oak *(Quercus velutina)* and white oak *(Quercus alba)* trees over a carpet of Pennsylvania sedge *(Carex pensylvanica)* and wildflowers. Look for round-lobed hepatica *(Hepatica americana)*, culver's root *(Veronicastrum virginicum)* and yellow pimpernel *(Taenidia integerrima)*, the only Michigan member of the carrot family with yellow flowers and smooth–edged leaves. Two uncommon low shrubs found here are hillside blueberry *(Vaccinium pallidum)*, and prairie willow *(Salix humulis)*, whose flowers appear before the leaves in early spring.

The **mesic forest** community has a nice display of spring flowers. If you can't get across the pond to visit the site, you can still enjoy the white blooms of serviceberry (*Amelanchier*) from the pull–off on Huron River Drive. Also look for the wild crab apple (*Malus coronaria*), a beautiful native apple tree of upland sites that is uncommon in the Ann Arbor area. In the fall, look for the purple aster-like flowers of Missouri ironweed (*Vernonia missurica*).

In the middle of this mesic forest is another small permanent pond which contains fish, and thus is not a good habitat for salamanders, who have better success reproducing in seasonally dry ponds or pools where fish are absent. The perimeter of this small pond contains some plants more typical of an emergent marsh, such as cat–tails. Canada geese have been seen nesting at the edge of this protected site.

Both the mesic and dry forests were burned by trained NAP staff and volunteers in spring 1996 and 1997 to remove the exotic honeysuckle (*Lonicera*) and buckthorn (*Rhamnus*) shrubs which are invading the area. With cooperation from Conrail, we also burned into the railroad right–of–way, which contains a remnant strip of **dry prairie** containing some beautiful native wildflowers.

Common arrowhead (Sagittaria latifolia) is also found in South Pond Park. Note its characteristic arrow-shaped leaves which stick up out of the water, and which give the plant its name.

The **wet shrubland** south of Huron River Drive is the only place in the Ann Arbor park system where the native highbush cranberry (*Viburnum opulus var. americanum*) has been found. The **wet meadow** here is dominated by the invasive purple loosestrife (*Lythrum salicaria*) in the central and eastern portions, while the western end is populated mostly by native species of Joe–pye weed (*Eupatorium maculatum, E. purpureum*), asters (*Aster*), and a nice variety of sedges (*Carex, Scirpus*). The thick vegetation and unpredictable water depth make maneuvering in this area difficult and it is best observed from the road.

GLOSSARY

Disturbance — Natural or human alteration of a natural community. Lightning–caused fire, flooding, and wind–throw are examples of natural disturbance. Fire ignition, fire suppression, road construction, and the establishment of impervious surfaces are all human disturbances.

Exotic — A species introduced to southeast Michigan following European settlement in the 1700s.

Impoundment — The body of water created by a dam, also called a pond.

Invasive — Aggressive to the point of out-competing species that would otherwise occupy the same habitat.

NAP — The Natural Area Preservation Division of Ann Arbor's Parks and Recreation Department.

Native — A species found in southeast Michigan prior to European settlement in the 1700s.

Natural Community — a grouping of plants and animals commonly found living together

Non–native — See *exotic.*

Mesic — moist

Savanna — A type of dry forest containing scattered trees, allowing more sunlight to reach the ground than in other woodlands. The understory consists mainly of grasses and flowering plants.

Seep — A freshwater spring trickling from the side of a slope. Seeps are formed when rainwater percolating down through more porous soil hits an impermeable layer and follows that layer horizontally to the side of the slope.

RECOMMENDED READING

Geology

Dorr, John A. and Donald F. Eschman. 1970. *Geology of Michigan*. Ann Arbor: University of Michigan Press.

Farrand, W.R. 1987. *The Glacial Lakes Around Michigan*. Ann Arbor: University of Michigan.

Hamlin, W.K. 1989. *The Earth's Dynamic Systems, Fifth Edition*. New York: Macmillan.

McGeary, D. And C. C. Plummer. 1992. *Physical Earth Revealed*. Dubuque, IA: Wm. C. Brown.

Skinner, B.J. and S.C. Porter. 1989. *The Dynamic Earth*. New York: John Wiley and Sons.

Plants

Barnes, Burton V. and Warren H. Wagner, Jr. 1981. *Michigan Trees*. Ann Arbor: University of Michigan Press.

Gleason, Henry A. and Arthur Cronquist. 1991. *Manual of Vascular Plants of Northeastern United States and Adjacent Canada, Second Edition*. Bronx, NY: The New York Botanical Garden.

Newcomb, Lawrence. 1977. *Newcomb's Wildflower Guide*. Boston: Little, Brown and Company.

Smith, Helen V. 1966. *Michigan Wildflowers*. Bloomfield Hills, MI: Cranbrook Institute of Science.

Voss, Edward G. 1972. *Michigan Flora. Part I, Gymnosperms and Monocots*. Bloomfield Hills, MI: Cranbrook Institute of Science.

Voss, Edward G. 1985. *Michigan Flora. Part II, Dicots (Saururaceae - Cornaceae)*. Bloomfield Hills, MI: Cranbrook Institute of Science.

Voss, Edward G. 1996. *Michigan Flora. Part III, Dicots (Pyrolaceae-Compositae)*. Bloomfield Hills, MI: Cranbrook Institute of Science.

Damselflies and dragonflies

Carpenter, Virginia. 1991. *Dragonflies and Damselflies of Cape Cod*. Brewster, MA: Cape Cod Museum of Natural History.

Dunkle, Sidney W. 1989. *Dragonflies of the Florida Peninsula, Bermuda, and the Bahamas*. Gainesville, FL: Scientific Publishers.

Kormondy, Edward J. 1958. *Catalogue of the Odonata of Michigan*. Miscellaneous Publications, Museum of Zoology, University of Michigan, No. 104.

Butterflies

Feltwell, John. 1993. *The Illustrated Encyclopedia of Butterflies.* London: Blandford.

Glassberg, Jeffrey. 1993. *Butterflies Through Binoculars.* New York: Oxford University Press.

Heitzman, J.R. and Joan E. Heitzman. 1987. *Butterflies and Moths of Missouri.* Jefferson City, MO: Missouri Department of Conservation.

Opler, Paul A. and Vichai Malikul. 1992. *Eastern Butterflies.* Boston: Houghton Mifflin Co.

Pyle, Robert M. 1981. *The Audubon Society Field Guide to North American Butterflies.* New York: Alfred A. Knopf.

Scott, James A. 1986. *The Butterflies of North America.* Stanford, CA: Stanford University Press.

Tilden, J.W. and Arthur C. Smith. 1986. *Western Butterflies.* Boston: Houghton Mifflin Co.

Wright, Amy B. 1993. *Peterson First Guides: Caterpillars.* Boston: Houghton Mifflin Co.

Xerces Society/Smithsonian Institution. 1990. *Butterfly Gardening.* San Francisco: Sierra Club Books.

Fish

Michigan Department of Natural Resources. 1995. *Huron River Assessment Appendix.* Hay—Chmielewski, et al.

Rosenthal, M. 1989. *North American Freshwater Fish.* New York: Macmillan.

Smith, Gerald. 1990. *Guide to Nongame Fishes of Michigan.* Ann Arbor: University of Michigan.

Zimm, Herbert and H. Schumaker. 1987. *Golden Guides Fish.* New York: Western Publishing.

Amphibians and Reptiles

Conant, R. and J.T. Collins. 1991. *A Field Guide to Reptiles and Amphibians: Eastern and Central North America, Third Edition.* Boston: Houghton Mifflin Co.

Harding, James H. And J. Alan Holman. 1990. *Michigan Turtles and Lizards: A Field Guide and Pocket Reference.* East Lansing: Michigan State University Cooperative Extension Service.

Harding, James H. and J. Alan Holman. 1992. *Michigan Frogs, Toads, and Salamanders: A Field Guide and Pocket Reference.* East Lansing: Michigan State University Cooperative Extension Service.

Holman, Alan J., James H. Harding, Marvin M. Hensley, and Glenn R. Dudderar. 1989. *Michigan Snakes: A Field Guide and Pocket Reference.* East Lansing: Michigan State University Cooperative Extension Service.

Birds

Harrison, Hal H. 1975. *A Field Guide to Bird Nests.* Boston: Houghton Mifflin Co.

Kielb, Michael A., John M. Swales, and Richard A. Wolinski. 1992. *Birds of Washtenaw County, Michigan.* Ann Arbor: University of Michigan Press.

McPeek, Gail A. and Raymond J. Adams (editors.) 1994. *The Birds of Michigan.* Bloomington: Indiana University Press.

National Geographic Society. 1983. *Birds of North America.* Washington, DC: National Geographic Society.

Payne, Robert B. 1983. *A Distributional Checklist of the Birds of Michigan.* Miscellaneous Publications, Museum of Zoology, University of Michigan, No. 164.

Peterson, Roger Tory. 1980. *A Field Guide to the Birds East of the Rockies.* Boston: Houghton Mifflin Co.

Robbins, Chandler S., Bertal Bruun, and Herbert S. Zim. 1983. *Birds of North America.* New York: Golden Press.

Mammals

Baker, Rollin H. 1983. *Michigan Mammals.* East Lansing: Michigan State University Press.

Burt, William H. 1957. *Mammals of the Great Lakes Region.* Ann Arbor: University of Michigan Press.

Burt, William H. and Richard P. Grossenheider. 1976. *A Field Guide to the Mammals.* Boston: Houghton Mifflin Co.

Kurta, Allen. 1995. *Mammals of the Great Lakes Region.* Ann Arbor: University of Michigan Press.

Murie, Olaus. 1954. *A Field Guide to Animal Tracks.* Boston: Houghton Mifflin Co.

Wild geranium *(Geranium maculatum)* is a common woodland wildflower. In late spring to early summer the pink–lavender flowers of this plant add color to the woodland.